ALTERNATIVE
CAREERS
IN SCIENCE

• • • • • • • • • •

Leaving the Ivory Tower

ALTERNATIVE CAREERS IN SCIENCE

· · · · · · · · · · · · · · · · ·

Leaving the Ivory Tower

Edited by

Cynthia Robbins-Roth

ACADEMIC PRESS

San Diego   London   Boston   New York   Sydney   Tokyo   Toronto

Front cover photograph © 1993, ColorBytes, used with permission.

Background map is copyrighted by the New York City Transit Authority and
is used with permission

Back cover photograph by Mark Richards.

This book is printed on acid-free paper. ⊗

Academic Press
a division of Harcourt Brace & Company
525 B Street, Suite 1900, San Diego, California 92101-4495, USA
http://www.apnet.com

Academic Press Limited
24-28 Oval Road, London NW1 7DX, UK
http://www.hbuk.co.uk/ap/

Library of Congress Card Catalog Number: 97-80573

International Standard Book Number: 0-12-589375-2

PRINTED IN THE UNITED STATES OF AMERICA
98 99 00 01 02 03 BB 9 8 7 6 5 4 3 2

CONTENTS

3. SCIENCE WRITING: COMMUNICATING WITH THE MASSES   **21**

Sue Goetinck

4. SNAKES AND LADDERS: THE GAME OF PUBLISHING   **33**

Anita Crafts-Lighty

5. BROADCAST SCIENCE JOURNALISM: WORKING IN TELEVISION, CABLE, RADIO, OR ELECTRONIC MEDIA   49

Eliene Augenbraun
Karin Vergoth

6. PITUITARIES TO PINSTRIPES: A PATH TO VENTURE CAPITAL   63

Andrea Weisman Tobias

7. HOW I BECAME AN ANALYST: SCIENCE-BASED INVESTMENT ADVISOR   71

Mary Ann Gray

PREFACE

When I first entered the classic path to becoming an academic scientist, I never dreamed that I would stray so far from that path. I entered worlds that I had never even considered—business development, journalism, and running a small business. And yet, even after all these years away from the laboratory bench, I can truthfully say that science remains the motivating force behind all my activities—my love of discovery, of learning about new areas of science. The only real change is that my focus now is bringing that science into the rest of the world.

When I began this journey, there were few road signs to follow—biological scientists did not leave the lab. I basically picked up my machete and forged a path from the bench to the board room. Today, more than a decade later, I have the chance to be a mentor to others who are making the same trip. I am deluged with phone calls from graduate students, tenured professors, and industry scientists who, for a wide range of reasons, are looking for clues to life outside the lab. I hope that this book inspires them to make the leap.

This book is dedicated to all of those who helped me in my evolution, and most especially to Dr. Steve Bennett, the M.D./Ph.D. who first led me into the world of venture capital; Brook Byers, who helped build an entire new industry around science; Dr. Stelios Papadopoulos, an academic scientist who helped finance that new industry; Joan O'C. Hamilton, who taught me the most important things about journalism; Dr. Steve Spencer, who watched way too many episodes of *The A Team* with me as I tortured myself over leaving the lab; Dr. Carol Hall, my long-time partner in BioVenture

Consultants and contributor to this book, who forced me to see the beauty in finance and to appreciate the synergy of two diverse minds working together; my parents, who have encouraged me every step of the way; and most of all my husband, Robert Roth, an M.D./Ph.D. whose love and support as I headed into uncharted territory gave me the courage to go for it, and who is just beginning on that same journey himself.

Cynthia Robbins-Roth

CONTRIBUTORS

Numbers in parenthesis indicate the pages on which the author's contributions begin.

DAVID APPLEGATE (207), American Geological Institute, Alexandria, Virginia 22302

ELIENE AUGENBRAUN (49), ScienCentral, Washington DC 20016

M. J. FINLEY AUSTIN (225), Merck Genome Research Institute, Lansdale, Pennsylvania 19446

RON COHEN (101), Acorda Therapeutics, Inc., New York, New York 10019

ANITA CRAFTS-LIGHTY (33), Biotechnology Publications, PJB Publications Ltd., Richmond Surrey TW10 6UA, United Kingdom

MARK D. DIBNER (247), Institute for Biotechnology Information, Research Triangle Park, North Carolina 27709

PETER DRAKE (85), Vector Securities International, Deerfield, Illinois 60015

SUE GOETINCK (21), The Dallas Morning News, Dallas, Texas 75214

MARY ANN GRAY (71), SBC Warburg Dillon Read, Inc., New York, New York 10022

GENEVIEVE M. HADDAD (235), Air Force Office of Scientific Research, Bolling AFB, Washington DC 20332

CAROL HALL (109), BioVenture Consultants, Chestnut Hill, Massachusetts 02167

P. W. "BO" HAMMER (201), American Institute of Physics, College Park, Maryland, 20740

BENTE HANSEN (189), DHR International, Inc., San Diego, California 92130

ERIN HALL MEADE (175), Lockheed Laboratories, San Jose, California 95132

ELIZABETH MOYER (119), Kinetek Pharmaceuticals, Inc., Vancouver, British Columbia, Canada V6P 6P2

RON PEPIN (93), External Science and Technology, Bristol-Meyers Squibb, Princeton, New Jersey 08543

CLAYTON R. RANDALL (11), PE Applied Biosystems, Foster City, California 94404

CYNTHIA ROBBINS-ROTH (1), BioVenture Consultants/BioVenture Publishing, San Mateo, California 94403

TONY RUSSO (161), Noonan/Russo Communications, New York, New York 10001

PAMELA SHERWOOD (133), Bozicevic and Reed LLP, Palo Alto, California 94301

KATIE M. SMITH (141), Gen-Probe, Inc., San Diego, California 92121

SUSAN L. STODDARD (151), Mayo Foundation, Rochester, Minnesota 55905

ANDREA WEISMAN TOBIAS (63), Abingworth Management Ltd., London SW1A 1HA, United Kingdom

KARIN VERGOTH (49), Science Friday, Washington DC 20016

chapter

1

· · · · · · · · · ·

A SCIENTIST GONE BAD:

Or How I Went from the Bench to the Board Room

· · · · · · · · · · · · · · · · ·

Cynthia Robbins-Roth, Ph.D.

Principal, BioVenture Consultants and Editor-in-Chief, BioVenture Publishing

It all began so innocently—back in 1984, I was happily running gels and killing tumors in mice; one year later, I was wearing grown-up clothes and hanging out with vice presidents and chief executive officers.

Since that time, I founded *BioVenture View*, a monthly biotech industry newsletter, and *Biopeople Magazine*; I was the founding editor of the first daily biotech fax newspaper, *BioWorld*; and I started a consulting business with another former scientist, providing business, technology, and financial consulting to startup biotech firms and multinational pharmaceutical companies. While I talk, write, and help build companies around science, I haven't done a hands-on experiment since 1984.

And I couldn't be happier.

This completely unplanned-for transition has led me into more opportunities to learn about new science, to spend time with world-class researchers pushing back new frontiers, and to communicate the excitement and promise of that technology with the nonscientific world. And what other biochemist can claim to have been quoted in that respected scientific journal, *Town & Country*?

When I crossed that line from scientist to "suit" person, there were very few examples for me to study. Researchers in the biological sciences were just starting to believe that it might be okay to leave academia and go into biotech companies. I literally did not know a single person who had made the transition—and I had no idea what a scientist out of the lab could actually do! I had spent my entire science career immersed in a very rarefied environment, surrounded by other biomedical researchers who saw the move to company-based science as the most unusual career move you could make.

This book gives you the insider's story on 22 different ways to put that scientific training to good use away from the lab bench, and for the most part, outside of academia. Each of our authors took an unexpected detour into worlds that were previously unimagined during their early science training. And while each of these jobs took the authors far from their original paths, the key to their success and enjoyment of the new task at hand was the underpinning of science that was crucial to excelling.

Right now, you probably can't imagine why a science background can be valuable in being a stock analyst, a publisher, an entrepreneurial chief executive officer, a venture capitalist, or a government policy expert. But, as you will learn from these personal stories, it is the science that taught our authors to think analytically, to structure an approach to new areas, to forge ahead into new territory without fear.

Even though all of us involved with this book, from the authors to the commissioning editor, are no longer in the lab, it was our science training that made us great at what we do today. The world is full of MBAs who long to enter the burgeoning biotech field, but who just can't master the intricacies of the technology sufficiently to be valuable to their companies or to their investors; patent lawyers who struggle with patent applications because they can't fully grasp the prior art in the scientific literature; information providers who don't understand that information they sell, and thus have a difficult time telling the difference between crucial and just interesting data.

Don't let anyone tell you that science is a dead end, now that becoming a full-tenured professor seems out of reach. And don't believe anyone who tells you that it is a waste of time to pursue a science education unless you plan to stay in the lab. There is a whole, wide universe out there, just waiting for you to explore it.

So How Did This Happen?

Back when I was in seventh grade, it came to me—I wanted to be a scientist when I grew up. The teacher was showing us how to drip acid on a piece of rock to discover if it was limestone. This simple-minded experiment had a huge impact on me, the idea that you could do experiments to figure out something that you didn't know already, that you could query the universe! This appealed to me immensely, probably in part because I had already developed a slight problem with authority figures and I really liked the idea that you could find out answers independently.

While the specific field of interest evolved over time, the basic drive toward lab work never changed. At Bates College in Lewiston, Maine, my biochemistry focus shifted a bit when I took my first immunology course, taught by a young scientist fresh out of his postdoc at Yale. Immunology was just on the verge of converting from a phenomenology (okay, let's stick this heterogeneous gamish into the bunny and see what happens!) into a realm where a protein biochemist could have some fun and learn some cool new stuff! My teacher was the first to let me into the wonderful of hands-on science—I was in love.

I moved to Texas for my graduate work in the lab of Dr. Benjamin Papermaster, where the focus was on applying the tools of biochemistry to purifying and characterizing the proteins that carried the messages (kill that tumor, knock off the virally infected cell!) for the immune system. I was intrigued by the idea that we could use our work to find a way to provide cancer patients with the immune factors their own systems were not making, moving away from the incredibly toxic chemotherapy drugs that were the only pharmaceuticals available at the time.

It was also increasingly clear that academia was not for me—I couldn't stand the thought of teaching one more medical student lab course. I started interviewing at pharmaceutical companies, but I was discouraged by their apparent propensity for hiring only middle-aged white guys as scientists (not to mention the white shirts and ties in the lab). Those companies seemed too conservative for me, and my little problem with authority figures had not gone away.

In 1980, I was in the midst of a postdoc in the interferon lab of Dr. Howard Johnson when I got a phone call from a scientist from a newly formed company—Genentech, Inc., in South San Francisco, California. While I had no idea what the heck this "biotechnology" industry was, my ears perked up when he said that the company was only 2 years old and had chosen cytokines—my area of interest—as an initial research focus. We agreed to meet in Paris, at the week-long International Immunology Congress. To this day, I am convinced that it was my ability to order him his first full meal in France that got me the job, as much as it was my research.

In spite of howls of "traitor" from my academic colleagues, Genentech turned out to be exactly what I was looking for, in many ways. The labs were packed with young postdocs and new Ph.D.'s. with no commercial experience, along with just about every piece of equipment you can imagine. The corporate environment was very entrepreneurial—Genentech was one of the first biotech companies formed, and it changed the ground rules for doing science in a corporate setting. Dress codes were nonexistent, scientists kept whatever schedules they wanted (being ex-postdocs, we all worked 18-hour days, at least 6 days a week), and we didn't have to write grants or teach medical students! I was working with some of the best scientists in a broad range of disciplines—protein chemistry, immunology, tumor biology, molecular biology, X-ray crystallography, amino acid sequencing, cell assay development, and so on. I was in heaven.

It was at Genentech that my "jack of all trades, master of none" personality, first nurtured in Johnson's lab, really came into play. While I was supposed to be focused exclusively on design of assay systems and purification schemes for certain cytokines, I spent a lot of time wandering the halls and learning how to do amino acid composition and sequencing, RNA purification (and why you really don't want phenol on your hands), and some really hard-core protein biochemistry. I learned about the problems in designing productive animal studies, the challenges of clinical development of complex biologicals, and the even bigger challenges of growing a young entrepreneurial business from 75 people who all knew each other, to 150 people, to 300 people and more.

During my Genentech stint, I spent time as a project team leader. The phrase "herding cats" springs to mind when I remember what it was like to get a group of aggressive, competitive scientists from different departments to quit bickering and start cooperating so that the project could move forward. This experience convinced me that people management skills were crucial to a successful business—and it also convinced me that I needed to improve those skills.

The other change in my thought process was the realization that science for science's sake wasn't all that satisfying to me—I wanted my work to contribute to developing a new therapeutic that could help patients. I wanted to know how the company decided which science projects would make the best products—what issues, besides science, had to be considered?

As luck would have it, my incredible ineptness at corporate politics and frustration with the "pushing on limp spaghetti" aspect of team-building in a nonteam environment propelled me out of the lab and into the best place to learn the answer to that question—business development.

I wanted desperately to leave Genentech and the constant battles, but I couldn't find a bench job in the area that would not be in conflict with my project at Genentech. I literally had no idea how to find a nonbench job. The only scientist I knew who had made the transition was a Ph.D. biochemist who became a patent lawyer—and law did not beckon to me at all. I started scanning the newspaper want ads and reading the classified ads in the back of *Science* and *Nature*. Months went by before I stumbled on an ad for "Advisor to the CEO" at a company I had never heard of, California Biotechnology, Inc. I had no idea what California Biotechnology, Inc., did, but what the heck, they were looking for a Ph.D. with biotech experience, and I certainly could give advice! (Of course, my shy and retiring personality might be a drawback . . . .)

I sent in a résumé, and got invited in for an interview with the CEO and the vice president of business development. It turned out that Cal Bio was another biotech firm, and the CEO wanted me to help analyze the huge number of ongoing science projects and help the management team determine which were great product opportunities and which were not.

The perfect job! I had to learn how to analyze science not just from the perspective of experimental design and data, but also through examining intellectual property issues, competition—not only from those in the same technology area, but also from other areas that might compete in the marketplace; I had to build a network of clinicians and learn what they saw as critical medical problems that required a novel therapeutic approach; I had to understand what it would take to develop such a therapeutic from bench to FDA approval and through into the marketplace; and I had to learn what other companies, both big pharma and biotech, were in that marketplace already. Luckily, I had a great mentor—Stefan Borg, who had a molecular biology background plus an MBA degree. Stefan taught me the basics of business development, and encouraged me on a daily basis.

I loved it! My science training in tracking down information and learning how pieces of data fit together to form a picture came in very handy, along with the ability to critically evaluate and analytically work through to a potentially unexpected answer. I absolutely loved working in such a broad range of disciplines. It was my job to track down experts in various fields and entice them into telling me everything I needed to know. In other words, I got paid to be educated in new areas by the experts—what a great concept!

I found that my scientific background and degree were almost more important than any growing business sense—just mentioning that Ph.D. and the labs in which I had worked gave me instant credibility with wary scientists who were getting really tired of talking to nonscientists—the bankers, analysts, lawyers, and corporate executives who wanted to put their inventions to work.

As I started putting together my reports to the management team on these projects, I found that I had to communicate the key concepts and issues in language that they understood. If the CEO didn't grasp the gist of my recommendations, all that work was useless to the company, no matter how profound my analysis. This basic fact of life forced me to improve my writing skills, which consisted strictly of knowing how to write dry journal articles for people in my very narrow field. I found that I loved to write, and I loved to find ways to communicate to others my excitement about an area of science.

This led me to start an in-house newsletter—the *RR Report*—to keep management informed about the activities of the competing biotech and pharmaceutical companies, and to highlight interesting scientific papers and conference presentations that might be good projects for the company. I had to learn how to use computers and word processing programs to generate my reports and the newsletter—the departmental secretary had informed me that she "didn't work for girls." I learned how to build databases, because it was the only way to keep track of the information in my newsletters. While I whined about having to learn this stuff back then, it gave me the confidence to tackle computer hardware and software (it also taught me not to read the manual).

This little newsletter brought me to the attention of Dr. Steve Bennett, an M.D./Ph.D. from Stanford, who had traveled around the world for the World Health Organization. When he tried to retire, he was recruited to run the medical portfolio of G.T. Capital Management's Emerging Technologies fund. Steve was an important investor in Cal Bio and he decided that putting a Ph.D. in business development was a great idea. He introduced himself, and that was the beginning of a beautiful friendship. Steve also introduced me to Dr. Carol Hall, a scientist-turned-Wall-Street-analyst, who got me thinking about the role of public investors in the industry.

In 1986, when I found myself on the wrong side of company politics once again and decided that I just wasn't cut out to be an employee, I met with Steve to see if he had some good ideas. I had a job offer—a real promotion over my Cal Bio position—from another biotech firm but I really did not want to go to work for someone else. Steve's answer was, "So become a consultant! In fact, come work for my fund." And so I took another great leap—into self-employment and the wild world of venture capital. It was the strong support of my husband, still ensconced in the academic science world, that gave me the guts to try.

Once again, an unplanned move turned out to be the perfect place for me. The flexibility of working for myself was a blessing, since I had a 6-month-old son. I love working with the fund, which was heavily invested in both public and private biotech firms. Steve taught me to understand some of the issues and agendas that investors must face, especially fund managers and venture capitalists who have investors and limited partners who whine about return on investment. He also showed me just what a great investor can bring to a company besides money—he was an active participant on the boards of directors of several of his portfolio companies and he actively stayed in touch with others.

The 1980s were heady days for biotech venture investing, and through Steve, I met and worked with many of the key players. I admired the way this first generation of biotech venture capitalists was focused on building successful operating businesses around this emerging new technology. They had to be swift on their feet, as the rocky ride on Wall Street made financings a challenge and the long product development time frame made smart management critical for survival. This group taught me how to pay attention to a broad range of issues, and not just to get enamored of the sexy science.

I thrived on the chaos and uncertainty of consulting. I loved working on multiple projects simultaneously. I had the opportunity to interact with a huge number of very smart people (most of whom also had great senses of humor), and I started developing a powerful network of people in and around the industry. I learned quickly that you always try to treat people fairly, and always help out when you can—you never know when and where that person will show up again! The phrase "what goes around, comes around" is really true.

But even as I became immersed in the consulting world, I was starting a second business—an industry newsletter. After I had been working with Steve's firm for about a month, he suggested that the *RR Report*, which he missed, would make a great publication for the venture community that was investing in this emerging industry. He offered me free run of the G.T. offices to work on the newsletter (access to computers, printers, copiers, and the all-important postage machine) in return for my more-regular presence in the office. And so *BioVenture View*, the first newsletter to focus on the business of biotechnology, was born in mid-1986 at 3000 Sand Hill Road, the hotbed of biotech venture capital.

The consulting and publishing sides of my business were a great synergistic fit—my monthly writing fits improved my writing skills immensely, and the increasing flow of company news into my files gave me a growing database of information for the consulting business. The key was keeping an absolutely strict confidentiality in place—no consulting client ever found their confidential information on the pages of *BioVenture View*.

As the newsletter grew in circulation, spreading to the executives in the venture-backed firms and then to the wider biotech community, I was struggling to keep up with the business. I spent my evenings and weekends on the living room floor, stuffing and sealing envelopes full of newsletters or reminder notices or invoices. Carol Hall had reentered my life, as a constant commentator on the wonderful world of biotech finance (she was now an investment banker) and an occasional *BVV* author on financial topics. It was great having someone else whose view of science and business was so complementary to mine. By late 1988, she had quit her job at a leveraged buyout firm and joined BioVenture Consultants full-time as my partner.

When my second child was about 6 months old, another change came. I was recruited by David Bunnell, a well-known publisher in the computer world, to put together the first daily on-line news service—focused on the biotech industry. This made me an employee again, although at least this time I ran my group (the editorial room) and was part of the "Gang of Four" that managed the company. Bunnell's team had great experience in the publishing world, and together we created *BioWorld*.

I had to create a newsroom that could provide very specialized reporting, including copy editors (I had never been copy-edited!) and reporters, and I built a true dream team. We lured Joan O'C. Hamilton away from her long-time post as chief biotech reporter for *Business Week;* along with Ray Potter, the biotech reporter from the *San Jose Mercury News* and Chuck Lenatti, a longtime magazine/newsletter editor. We stole Carol Ezzel away from *Nature* to run our Washington, D.C., office and to set up a network of freelancers around the United States.

Running a daily newsroom was a complete change for me—the pace was outrageous, compared to the monthly crawl of *BioVenture View* (which I also was still producing), and I had never worked with professional journalists. But we built a great news service, which transformed reporting in the biotech world and made our industry suddenly very visible to more than just the hard-core insiders. Even the most nonscientific analyst was comfortable reading our daily faxed version of *BioWorld*, which was designed to resemble a newspaper.

While daily news can be very addictive, I was getting worn out from the relentless pace and I was also chaffing once again at being an employee. So, in early 1991, I left *BioWorld* and took *BioVenture View* independent once again. Carol and I revved up the consulting business, which had been on the back burner while I ran the newsroom and Carol had her second child. Our client base evolved away from the venture capitalists and toward working directly with early-stage biotech firms that needed help. We wrote business plans, we helped with restarts of companies whose initial strategy didn't work out, we assisted with public and private financings, we worked on finding corporate partners to help companies develop and commercialize their products, and we generally had a great time. Essentially, we got paid the big bucks to learn cool new science and boss around people in suits.

As the consulting business grew up, the networks that Carol and I had nurtured all those years paid off. They brought us great consulting projects and access to the experts whose input we and our clients needed. In many ways, we act as biotech yentas, matchmakers who look at science, financial issues, and management teams to help bring together the critical mass for the successful growth of the companies. We work with brand-new companies, established multinational pharmaceutical firms, companies in Canada, Asia, and Europe, as well as in the United States.

As time has gone by, we are spending an increasing amount of time giving invited talks at industry conferences—talks about the overall biotech industry, about trends in financings and corporate partnering, about our annual *Women in Biotechnology* survey that tracks women board members and members of top management, and about moving from science into new careers.

Carol and I are both the beneficiaries of mentors who helped us find our way in new universes, and we field calls every week from students, professors, and industry scientists who are looking for some suggestions on how to investigate the possibilities outside the lab. The increasing pressure on federal funding of research and the increasing number of Ph.D.'s in many scientific disciplines has meant that the demand for academic jobs far outstrips the supply. More and more scientists are forced to consider alternatives at an earlier and earlier stage in their careers.

Even though I don't run gels and columns anymore (and I'm not sure I can still do cardiac puncture on a rat), I very strongly believe that the key to my contribution to our clients lies in my hard-core, hands-on science training. While I have spent the past 15 years learning many other disciplines, it is that core experience that informs how I think and analyze, how I bring together apparently disparate pieces of information. The driving force behind my enthusiasm for my work is the love of the science that lies behind it all.

This book is full of personal stories from individuals who all share a strong scientific background—they are all Ph.D.'s. or M.D.'s. While they ended up in an amazingly diverse collection of jobs, there are certain recurring themes that you will find in all of their stories. The first is the willingness to take a chance, to give serendipity an opportunity to work. The second is an amazingly strong streak of self-confidence, the willingness to chance falling flat on their faces. This doesn't mean that everything went smoothly for our authors. On the contrary, you'll find that most of them faced some real adversity—but in fact this very impediment to their expected path is what flung them into areas that they have come to love. All of these authors have learned how to build strong interpersonal relationships, even those of us who really have problems taking orders! The ability to create a team—of mentors, of staff with new disciplines—allows you to make almost any transition necessary.

I encourage all of you reading this book to pursue these far-flung opportunities, and find some of your own. Oh, and remember that nice young postdoc who first let me loose on an unsuspecting laboratory? He is now wearing a suit as Assistant Director of the Office of Science and Technology at Duke University, building alliances between the scientists and clinicians on campus and the biopharmaceutical industry. His broad scientific knowledge and ability to understand the issues affecting both sides of those negotiations is the key to his success.

So take off your lab coat and check out this new universe. You never know where you might end up.

chapter

2

TECHNICAL WRITING:

Making Sense out of Manuals

Clayton R. Randall, Ph.D.

Senior Technical Writer, PE Applied Biosystems

WHAT IS A TECHNICAL WRITER?

The broadest definition of a technical writer is someone who compiles large quantities of information into a useful, easily digested form for a specific audience. A journalist for *Scientific American* is a technical writer; the person who writes instruction manuals for *Mathematica* or *Doom II* is a technical writer. So are you, if you've written scientific papers (although the "easily digested" part doesn't apply as much).

In the San Francisco Bay Area where I live, most technical writers write manuals and online help files documenting computer software and hardware, although more jobs are becoming available in other fields such as biotechnology.

I work for the life science division of a large analytical instrument company that is the world's leading provider of DNA synthesis and analysis equipment and the reagent kits that are used on the equipment. I write instruction manuals for the reagent kits, which have specific applications such as DNA sequencing, human identification from forensic evidence, and food testing for *Salmonella*.

WHY TECH WRITING IS A GREAT MOVE

Here are some reasons why technical writing is a good career transition for scientists:

- You don't have to stop being a scientist. You still have daily contact with scientists and discuss their work with them. Besides, it's impossible to stop being a scientist. It's in the blood and it's how we were trained to think. I often refer to myself as a "recovering scientist," because I left research but will always be a scientist.

- You have instant credibility with "subject-matter experts." Because you're also a scientist, the scientists and engineers with whom you work think automatically that you know what you're doing. They don't think that way about all tech writers, which is a downside of the field. It's up to you to maintain that credibility by the way in which you perform your job.

- You don't have to do lab work. The endless repetition of experiments is one reason many of us left research. As a writer, you get to learn about new scientific methods and new software without having to do the grunt work of repetition and method validation. The thrill of learning is why I became a scientist, not the drudgery that can go along with it. When a tech writing project is done, you go on to the next project. You don't have to keep doing the same thing for years on end.

- You can still do some lab work if you want. The best way to learn about a new product is to test it, whether it's a reagent kit or piece of software. In fact, the product team will respect you greatly if you take the product for a test drive.

- You resemble the customers. As a biotechnology writer, I have a lot in common with the end users of my company's products, who are research scientists and technicians. One of my co-workers, another Ph.D. chemist, got her first job because she used certain computational software extensively. The software company needed a new manual at the same time her postdoc funding was running out. Because you resemble the customers, you understand their concerns and can write about what they want to know.

- Your will have a less-steep learning curve than will most writers. You already know a lot of science and you have the background to learn more science easily. And don't forget that you're more computer literate than 95% of the population. Scientists use computers to collect their data, analyze it, and write up the results.

- Your work ethic will be appreciated. Most scientists work hard, are inquisitive, and pay close attention to detail. These are all useful qualities in technical writers. We need to go out and learn as much as possible about our subject, then make sure that we write carefully and accurately.

Overall, scientists and technical writing are a good fit. Your technical background, willingness to learn, and meticulousness will serve you well as a tech writer.

COMMON PERSONALITY TRAITS OF TECH WRITERS

Lifelong learning is the most important trait of a technical writer. You should be interested in almost everything and willing to learn as much as you can about it. If you're excited about your subject, your writing will be interesting as well.

You have to like writing. I realized that one of the things I consistently enjoyed about science was seeing my work in print. I also enjoyed explaining my work to my nonscientist friends, who always said that I was good at making science understandable. I discovered that there were people, called technical writers, who did this for a living. The idea of being paid good money to write was appealing.

You must enjoy working with people. Most of the information you use to write documentation comes from interviewing subject-matter experts singly or in groups. Even if you are given a written draft of a procedure, you will still have to talk with its author to clarify and complete it. The networks you establish by trading information and favors are very important in getting your job done (more on networking later). It's not necessary to be a complete extrovert, but you have to be able to talk to many different kinds of people. Take an interest in them and learn about them, too. It's part of what makes the job fun.

Pursue knowledge actively and be tenacious. You have to go out and find what you need, rather than waiting for it to come to you and to be dropped over your cubicle wall. If you don't get what you need the first time, go back and bother the necessary people again and again until they give you what you want just to get rid of you. They have their priorities, but so do you.

Pay close attention to detail, but don't be consumed by it. Your grammar and spelling must be flawless and your content as accurate as possible, but don't try to write a perfect document. That takes too much time, and the product has to get out the door to make money so that everyone gets paid.

The ability to handle multiple tasks and projects simultaneously is crucial. It's common to work on several projects at once, each with its own issues and deadlines. Even if you work on only one or two large projects at once, each can have many chapters and bits of information to juggle.

Finally, it is actually good that I have a short attention span. I'm interested in almost everything, learn quickly, and get bored quickly. In grad school, I had trouble staying focused on the same project for years. Now, I work on several projects at once, and finish 10 or 15 projects per year, ranging from 30-page reagent kit protocols to 250-page user's manuals. Mostly, I write shorter documents. When something is done, I move on.

Some tech writers prefer the opposite way. They like to work on the same large project for 6 months, a year, or longer. What would be anathema to me works for them. Still other tech writers fall somewhere between the two extremes and work on a variety of projects.

A Typical Day at Work

I usually get in between 9:00 and 9:30 A.M. I turn on my computer and while it's booting up, check my voicemail to see what crises have arisen since I left the night before. If I have no voicemail, I get a cup of coffee, the first of two or three during the day. Once the computer's up, I check my e-mail, then spend some time talking with my co-workers, especially if it's Monday and they have good stories from the weekend.

After returning phone calls and putting out fires, I spend most of the morning writing and editing, unless I need to get a manual to the printer or I have a meeting.

Writers in my department are responsible for all aspects of their documents, from planning to writing to getting the electronic deliverables (printable files) ready to send to the print vendor. I've also put some of my manuals on the company Web site.

Each project team usually meets weekly to discuss progress, which means I have four to seven meetings each week, not including impromptu meetings called to review drafts of manuals or to handle sudden research or manufacturing issues. Projects, such as a new instrument or reagent kit, are accomplished by interdisciplinary teams composed of people from marketing, research & development, process development, manufacturing, technical communications, and so on. There are also staff meetings, where a department such as technical communications gets together to discuss what happened and what its people did since the last meeting.

After the morning writing and/or meetings, there's lunch. I usually go offsite with a few co-workers or one or two friends from other departments (we have a cafeteria, but I don't like the food). I met most of my friends in other departments by working on project teams with them. If I've just finished a project, I reward myself by going out for sushi. Celebrating accomplishments is important in any job.

After lunch, I check voicemail and e-mail again. (On a slow day, I check e-mail about 10 times.) The afternoon is spent like the morning: writing, editing, or in meetings. I take some time to walk around and check with people in the various project teams if I need information from them or if I have information to share with them. I also return voicemail from people looking for a manual or for information that I have or that I might know where to find.

I enjoy networking, and I find that if I go out of my way to help other people, they'll do the same for me. Because I work on many projects across the company, I know more of what's going on than do most people who work in only one department. I know about new projects before they're dropped on my desk.

After 5:00 P.M., our department gets quieter as people go home. I work for another hour or so, check the e-mail one last time, then leave around 6:00 or 6:30. We don't have set hours; we just have to get the work done.

WHAT I LIKE BEST ABOUT MY JOB

The best thing about my job is that I write creatively while helping people. They follow my instructions to use our company's reagent kits to test food for harmful bacteria, to find new genes, or to show which suspect could or could not have committed a crime. My work has very tangible results.

When I explain my job to my friends and relatives who aren't scientists, I tell them about the real-world applications of my work, adding that I "explain science to normal people." They immediately understand and approve. Try explaining the K_β X-ray fluorescence spectroscopy of metalloproteins (my postdoc work) and its utility (if any) to your relatives in less than 100 words. It doesn't work.

Technical writers help people do what they need to do. The corollary to this is that our work is read by many people. A friend of mine pointed out that my work is read and used by thousands or tens of thousands of scientists, more than any but the hottest papers in *Science* or *Nature*. Besides, I get to see my work in print 10 or 15 times a year.

On a more mundane level, I like being told that I've done a good job. That seems to happen more in industry than in academia, where you're told, "good work, but you need to repeat it and do five more studies like it."

What I Don't Like about My Job

Like any field, technical writing has its disadvantages. Some of them are common to other fields as well.

Repetitive stress injury (RSI) is the biggest job-related hazard for technical writers. Because we type and use the mouse (the most anti-ergonomic thing in existence other than high-heeled shoes and neckties) most of the day, our wrists and hands are particularly vulnerable. Eye strain and lower back strain can also occur. When working, take frequent breaks—at least once every half hour—especially if you're working a lot with a mouse. If your hand is often sore, you probably already have an RSI. If you have chronic pain, see your doctor.

Tech writers are also susceptible to stress, mainly because of deadlines and juggling too many projects at once. Frustration commonly occurs with project teams who think that your document is a much lower priority than you do and act accordingly. Frustration with project teams who think you're a glorified typist also occurs, although this occurs less often for scientifically trained writers. Sometimes there's frustration with your boss, who wants the document to be perfect when you just want to get it out the door.

Another source of frustration comes from paperwork. More and more companies are obtaining ISO 9000 certification, which states that the company follows certain written procedures consistently. ISO 9000 works well for manufacturing processes, but not as well for technical documentation, where it can complicate your job unnecessarily. The problem often is not the standards themselves, but how they're written. The people who write the standards seem to feel that because there must be standards, the standards must be complex. In this case procedures can get in the way of doing your job.

How Is This like Grad School and Postdoc Work?

Skills—Technical writing uses many of the same skills: analytical, scientific, and computer.

Challenges—Writers are constantly learning new scientific procedures and computer software.

Networking—Writers have to make contacts to find out the information they need to do their work, to find out the latest advances in their company's field or in writing, and to find out where the goods jobs are. This includes attending scientific or technical writing conferences.

How Is This Different from Grad School and Postdoc Work? (The Bottom Line)

Salary—According to the Society for Technical Communication (sort of like the AAAS for tech writers), the average technical writer in the United States makes from $45,000 to $50,000 per year. In the Bay Area, which has the most technical writers and technical writing positions, it's closer to $55,000. Starting salaries here are in the mid-$40,000s, which is comparable to the average starting salaries for Ph.D. research scientists without postdoctoral experience.

Independent contractors and consultants (freelancers) can make much more than that (hourly rates start at $50 or $60 for experienced writers), but they usually have to pay their own benefits. In general, an experienced technical writer's salary will plateau around $70,000 unless that writer becomes a manager or a freelancer.

Job stability—Technical writing is an expanding field with very good job prospects in many areas of the United States and Canada. It is also a very portable field: you can write almost anywhere that has phone service. Telecommuting is very common.

Better hours—Most writers work from 9 to 5, give or take half an hour to an hour, and never on weekends unless a deadline makes it absolutely necessary.

Equipment—Each writer has his or her own computer, usually a high-end Macintosh or Windows machine with a large monitor. Our department has two high-speed printers and a scanner. If we're writing about a new machine that the company plans to release, we can borrow a machine if it's small or we can use one in a lab if it's large. If we're writing about new software, we get a copy of it to play with.

Interpersonal skills—People skills are more important to a tech writer than to a research scientist, because we get information from other people, not from running experiments.

How to Become a Tech Writer

This is what worked for me and for other scientists who have successfully made the transition into technical writing.

- Rework your résumé. Most résumés of grad students and postdocs start with educational background and end with a list of publications with titles that are incomprehensible to anyone outside of the applicant's own subfield. This says "SCIENTIST" in large, bold type.

Instead, create a functional (skills-based) résumé. Start with a concise job objective, such as "to combine my scientific background and writing skills in a challenging position." List your skills and detail your experience, concentrating on writing and computers. Give brief summaries of your employment experience and education and tell how many publications you have. Your résumé should be no longer than one or two pages.

- Get some experience. Submit freelance science articles to your local paper. Volunteer to write for or edit the newsletter of the local chapter of the scientific society to which you belong. You need to build up a "portfolio," a body of your work that isn't made up only of scientific publications (although successfully funded grant proposals are a plus). Volunteer work is also a good line on a résumé.

- Learn a word processing or desktop publishing program. The more tools of the trade you know, the better. Start with the word-processing programs Word or WordPerfect, which most research groups use to write their papers. If you're serious about tech writing, get a copy of Framemaker, PageMaker, or Quark Xpress, which are desktop publishing programs. Most college bookstores have academic discounts on software. FrameMaker is the most expensive but most useful of the three, and this makes an excellent line on a résumé. Besides, you'll need to know one of these programs for your technical writing classes.

- Take technical writing classes. It helps to have a credential on your résumé that doesn't say "SCIENTIST." Writing classes are also helpful in learning to write for different audiences, and the classes will deprogram you from writing in the dry, passive-voice, scientific style. ("A 5-mL aliquot of dichloromethane was added to a round-bottom flask containing compound *1*. The mixture was then stirred at ambient temperature for an 8-hour period." Instead say, "Add 5 mL of dichloromethane to compound *1* in a round-bottom flask. Stir at room temperature for 8 hours.")

The classes are also great for networking. I found my job through a posting in one of my classes. I will probably find my next job through my contacts in other companies, many of whom I met through my classes.

Check with your local community college or state university campus for technical writing certificate programs. You can take classes part-time while you're still a grad student or postdoc. I started taking classes when I found out that the grant supporting my postdoc position wasn't renewed (I guess we didn't explain the utility of K_β fluorescence spectroscopy well enough). I was tired of research and was going to become a tech writer anyway, but the thought of impending unemployment gave me an incentive to leave research. I kept taking classes after my postdoc ended, and I was out of work for only 2 months when I found my internship.

- Find an internship. Working as an intern can teach you how to be a technical writer in the real world. These jobs are also easier to get than are full-time jobs if you have no experience. Be willing to accept relatively low pay ($8–$15 per hour) for 3 to 6 months. Internships can lead to a permanent position at the same company or at another company. Either way, you'll have marketable experience.

  After seeing a posting in one of my classes for an "aspiring tech writer" with scientific experience I started as an intern. I was lucky to be in the right place at the right time. After 3 weeks as an intern, there was an opening for a full-time writer at my company and I was hired. I started in February 1996, and was promoted to Senior Technical Writer in January 1997.

  Some recruiters will hire inexperienced writers as contractors. This is very useful experience, but be careful to read every word of the contract before you sign. Don't sign just anything because you think you can't find any other job. Also, if you do take a contract that isn't an internship, don't take less than $20 to $25 per hour for your first job.

- Join the Society for Technical Communication (STC). The STC, an international organization of about 15,000 technical writers, is based in Arlington, Virginia. It has chapters in every state in the United States. STC meetings are a great place to network. Each chapter also has a list of available jobs, most of them posted by members or local recruiters. STC contact information is listed at the end of this chapter under "Resources." Membership costs $95 per year ($40 for students in technical writing programs).

WHERE TO FIND TECH WRITING JOBS

Most technical writing jobs are not listed in the newspaper. If you don't have many contacts in the tech writing field, then the local STC chapter is the best place to begin your search. College placement offices and technical communication (or English, etc.) departments help their students find jobs.

Once you find a job, keep your skills and résumé current. Learn as much as you can about everything. Network as much as you can. You never know what will lead to your next job.

WHERE CAN YOU GO FROM TECHNICAL WRITING?

Most writers eventually become managers of other writers, or they become freelancers, that is, they work for themselves as contractors or consultants. They can also go into marketing communications (advertising and public relations work) or science journalism. Some writers prefer to remain as individual contributors at a company, but they move up in rank and pay by changing companies. This is the Silicon Valley way of getting promotions and raises.

Scientists have other career paths. Going back into the lab is an option, but it gets more difficult the longer you're out of the lab. In companies like mine, which are characterized by interdisciplinary teams and a relatively flat (non-hierarchical) organization, lateral moves are common. R&D scientists often switch research groups or move into process development or marketing. Most marketing people at my company came from the lab bench, as did the executives.

Scientists who are technical writers can also move into areas with more direct customer contact, such as marketing, technical support, sales, and field service. Our company also supports rotations, where you can work in a different area for a month or a few months. I learned a lot by working in our technical support department, taking customer phone calls for a few days, and I plan to do a rotation there for a month later this year.

Outside of industry, science journalism and science and technology policy are also possibilities, because people who work in public policy write many reports.

In short, there are very few limits to what you can do as a technical writer.

RESOURCES

The Society for Technical Communication (STC) is good for networking and for finding schools with technical writing programs. Membership costs $95 per year ($40 for students in technical writing programs). From its Web site, you can find links to every local chapter that has a Web site.

Society for Technical Communication
901 N. Stuart St, Suite 904
Arlington, VA 22203-1854
(703) 522-4114
http://www.stc-va.org (http://stc.org should work, too)

An excellent reference is *The Tech Writing Game,* by Janet Van Wicklen (Facts on File, New York, 1992), ISBN 0-8160-2607-6. It goes into much more depth about the profession than I can here. I highly recommend it to anyone who might want to become a tech writer.

Sue Goetinck, Ph.D.
Science Writer

Thinking about leaving the lab? Do you want sane hours and more variety in your job? Then science writing may be for you. I've been working as a science writer for *The Dallas Morning News* for 3 years, and I feel like I'm in college again. (Only now they're paying me!) Every week brings a new topic and I'm learning more about science then ever. If you're thinking about writing as a career, this chapter will give you a flavor of the field and tell you how to get the training you need to become gainfully employed.

SHOULD YOU BECOME A SCIENCE WRITER?

A personal ad for a scientist who would make a good science writer might go something like this: "BUS (bored, unfulfilled scientist), age and zodiac sign unimportant, likes to talk science but hates to do it. I would rather learn a little about many areas of science than spend the rest of my career focused on a single, narrow field. I crave variety. I need to feel like I've accomplished something when I go home for the day. I am willing to leave behind cumbersome, jargon-laden academic prose. I dream of getting published in a matter of a month, a week, even a few hours. Can you help?"

Absolutely! I have a Ph.D. in molecular genetics and I've been a science writer with *The Dallas Morning News* for 3 years. During that time, I've met dozens of other writers who have also left the lab. When we talk about how we got into the profession, the reasons are usually the same. Like scientists, we science writers are generally deeply interested in science. We're curious, we love to learn, and we want to make a difference.

But unlike most scientists, those who have become science writers are perfectly happy to observe. In fact, we are downright relieved that we never have to do another experiment. We're so happy being vultures—we can sit and wait until someone else has repeated an experiment 10 times, until they've worked out all the quirks, done all the controls and agonized over a publication. Then, we revel vicariously in their success, write an article about it, and get it in print in a fraction of the time. Doesn't this sound great?

A Challenging Field

For me, science writing has been as challenging as it's been satisfying. While I mainly cover biology for the *Morning News*, I've also had the opportunity to write about geology, astronomy, and particle physics. Those were some of the more difficult stories I've done, but what other job would allow me to spend hours on the phone quizzing some of the country's best geologists and physicists about their work? If I were still working as a biologist and wanted to expand my horizons, I'd have to settle for reading popular science articles that someone else had written.

Within my primary focus, or "beat," I've covered a huge variety of topics in life sciences—the biology behind mental illness and breast cancer, the cloning of sheep and the genetics of taste, heart development, high blood pressure, immunology, the Human Genome Project, and many other advances in genetics.

The job isn't easy, but it doesn't run me into the ground like research did. This, of course, is because I find reporting and writing a lot more fun. All that said, be forewarned that science writing is not a profession for scientists who just can't come up with a better alternative. If you venture this way, you'll work hard. Deadlines and editors can be tough. But you'll learn a lot, you'll reach real people, and (most days, anyway!) you'll go home knowing you've accomplished something.

WHO NEEDS SCIENCE WRITERS?

Lots of groups do. Science writers can be found almost anywhere science is done, and quite a few places where it's not. From my own unscientific observations, it seems that the most jobs are in a field known as public relations or public information. Virtually every university, medical center, hospital, or institution that does research has its own staff of reporters that keeps tabs on the happenings at the institution. Public information officers, or PIOs for short, write press releases that are distributed to the media, contribute to in-house publications and alumni magazines, and serve as a liaison between reporters and the institution's staff. They also monitor local and national media trends so they know which staff members might have something valuable to say to reporters about the latest hot issue.

Government research institutions, such as the National Institutes of Health or the Department of Energy also have PIOs, as do many companies with science-related products. And sometimes independent public relations firms represent these types of institutions.

Museums or aquariums are another option for science writers. These institutions need writers to publicize research and exhibits, as well as to help design and write signs for the exhibits themselves.

Other outlets include magazines and newspapers. (More about newspapers coming up, since that's what I do.) There are several popular science magazines—*Discover, Science News, New Scientist, BioScience*, and *Popular Science*, to name a few. Journals such as *Science* and *Nature* have their own reporters, and a lot of general news magazines such as *Time* and *Newsweek* also have specialized science writers. Health and trade magazines need writers with a technical background as well.

Some universities, such as Harvard, publish newsletters that summarize the latest advances and trends in a variety of medical specialties. And I've seen more and more jobs advertised lately for World Wide Web publications. Some of my colleagues have written for children's science books and TV shows. Writers who have been around for a while doing smaller pieces often venture into book writing. One science writer I know uses his writing skills to write speeches and memos for company executives. Another has helped edit an undergraduate biology textbook.

Many people have asked me about freelancing. This is an attractive option for someone who doesn't want to be tied down to a staff position. A word of caution, though. Most freelancers don't rely on freelance dollars as their sole source of income. Many have part-time jobs or a spouse who helps pay the bills. However, if you get some really good clients, it's possible to survive this way.

So to sum up, there are a lot of different types of a job for a writer who knows science. But first, you'll probably need a little training.

How Do I Become a Science Writer?

After finishing my Ph.D. at Washington University in St. Louis, I enrolled in a 9-month science writing program at the University of California, Santa Cruz. In that short time, I learned to write news stories, features, and essays. I got practice reporting and interviewing and did two 10-week internships at *The Californian*, a small paper in Salinas, Calif. I covered everything from community social programs to murders.

While not all science writers have formal training in journalism, I think it helps. Through the program at Santa Cruz, I met many professional writers and editors who gave me a good sense of the field. And I was able to get those invaluable "clips"—printed newspaper and newsletter articles that served as part of my portfolio when I applied for an internship at the *Morning News*.

There are several programs across the country that train science writers. There will be more about how to find these at the end of the chapter. If you can spare the time and the money, I would recommend applying. The American Association for the Advancement of Science also offers Mass Media internships every summer. The contacts you'll make and the clips you'll accumulate will help you get your foot in the door. But if taking one of these programs isn't an option, I offer a couple of tips.

First, do whatever it takes to get an article published! Even one clip can get you your next assignment. If you have to write an article for a small newspaper or a newsletter for free, do it! If nothing else, it will give you a flavor for the job. My first clip was a short article that was printed in one of the free community newspapers in St. Louis. I'm forever grateful to that paper's editor, because my modest clip helped me get into the writing program in Santa Cruz.

Second, read an introductory undergraduate journalism textbook. This type of book will introduce you to the basics. Even if you don't want to write for a newspaper, the advice in the book should help you get away from academic-style prose and move you toward a writing style that is more easily understood by lay people. Whatever type of writing is your goal, practicing the newspaper style will teach you to organize your thoughts, to write in a conversational style, and to cut out unnecessary words.

If you do attend a journalism or writing program, be prepared to work hard, and also to learn a whole new way of thinking. I learned more in my 9 months at Santa Cruz than I did in almost any 9-month stretch working on my dissertation.

What Is a Typical Work Day Like?

A typical day for me (and probably for most newspaper reporters) starts like this: I get to work around 9:30 A.M. I check my voice mail, my E-mail, and scan my snail-mail to see if there's anything pressing. Then, I read the *Morning News* and look through *The New York Times*. If I have a story in the paper, I check to see how it survived the late-night editing process. What happens next depends on the day of the week and what stories I happen to be working on. Early in the week, I scan "tip sheets," bulletins from the major science journals that summarize what is coming out in that week's issue. I order any papers that look like they might make a good story and save them to "pitch" at the weekly staff meeting. Then I might go to the local university to do an interview, do interviews over the phone, or work on a story. If I have a major story coming out in the weekly section, later in the week I'll discuss the text with my editor and the accompanying graphics with the artist. If there's a story that merits "daily" coverage (an article that has to run in the national or local section of the paper the next day), my work day might get a little hectic, especially if sources decide not to call back until an hour before deadline.

Sometimes news will break unexpectedly, such as when NASA scientists reported they had possibly found life on Mars, or when Scottish scientists announced that they had cloned a sheep. One morning, when I was still an intern, I lay half asleep and listened to National Public Radio announce that Al Gilman, a Dallas biologist, had won a Nobel prize. Needless to say, that was a busy day!

Big news doesn't break that often in science, especially compared to beats like crime or politics. So in between interviews and writing, I have time to read journals, press releases, and other publications. I travel to five or six scientific meetings every year to hear the latest, unpublished research and to meet sources face to face. Basically, I don't feel the need to rip my hair out more than, say, three or four times a year. My days generally end at 6:30 or 7:00 P.M.

It's Science, But Is It News?

One thing I had to adjust to quickly as a newspaper writer is the fact that the science that excites newspaper editors isn't always the same as the science that excites scientists. Ninety percent of the articles that appear in *Science* and *Nature*, for instance, are simply a waste of ink to any newspaper editor. And rightly so. If you're planning to venture into science writing for the public, remember that most people who read your stories, even if they are science buffs, do not care whether a particular enzyme needs magnesium to work. They do care, however, that a sheep has been cloned, and they care about some slightly less startling news too.

As the life sciences reporter at the *Morning News*, it's my responsibility to keep abreast of my beat well enough to know what makes a new development worthy of coverage. Most of the stories I do are my own ideas, but I also get assignments or suggestions from my editors. There is no formula; a lot depends on the publication you write for and the tastes of your editor. For example, my editor doesn't like to run profiles of scientists. He'd rather focus on their research. He won't accept a story on a research area that's been around for a few years; there has to be a more recent development. While "the last few years" is still considered new to scientists, to a newspaper, new usually means yesterday. That's an exaggeration, but stories do need a "news angle"—a reason to write that story now.

HOW DO YOU EXPLAIN DNA IN THREE SENTENCES?

You don't. One skill every science writer needs is the ability to explain complicated subjects in just a few words. That means you have to know what to put in and what to leave out. I've written 1500 word articles about genetics without mentioning DNA. "Genetic material," my favorite euphemism for DNA, does just fine in most cases. Sometimes, however, a story calls for a more detailed explanation of DNA. So genetic material becomes "a helical molecule that looks like two springs wound together." A mutation is a "misspelling." I can't say that "the gene is expressed only in the cone cells of the retina," but I can say that "the gene is turned on only in the cells of the eye that make color vision possible."

And so it goes. When you sit down and write an article about science, you shouldn't tell the reader any more than they need to know at that point in the story. Remember too, editors usually have a word limit. For daily newspaper articles, that means putting the meat at the top and saving details for later. Editors generally cut these stories from the bottom. Longer, feature-type articles may have more leeway. With practice, you'll get a feel for how much detail you can fit in a story of a given length.

DO SCIENTISTS LIKE REPORTERS?

Some do and some don't. Most people I call for stories are happy to talk to me. Generally, they are flattered by the attention, and they think it's important to communicate their research to the public. But occasionally, there are hurdles. The most prominent people in the field, for example, also tend to be the busiest. I once had to schedule a 15-minute phone interview a month in advance!

Other problems include sources who either never call back, or call back in a few days, well after the story has appeared in the paper. Rarer, but definitely out there, is the scientist who simply refuses to get on the phone with a reporter. There's not much to be done about them, so I don't waste my time trying to convince them.

The biggest worry scientists have is being misquoted. They are used to having control over everything else with their name on it—manuscripts, meeting abstracts, grant applications—and often get nervous when they can't see the newspaper article before it comes out. Our paper doesn't allow that, and I can usually reassure my sources by telling them that I don't want to be wrong in print any more than they do. If I have questions, I always call them back.

After the article appears, I always send copies to all my sources, whether I ended up quoting them or not. I've found that researchers appreciate this, and if they are the hard-to-reach type, they are more likely to return my calls in the future.

WILL ANYONE READ WHAT YOU WRITE?

If it's a well-written, interesting story, you can bet people will read it. I don't get daily feedback from my readers, but a couple of letters or phone calls a month isn't out of the ordinary. Most have been complimentary, but occasionally I hear from someone who's unhappy with something I've written, such as one article I did on evolution. A creationist didn't buy what I had written, and he sent me a nasty letter to that effect.

I've also received calls from people who oppose animal research. When I write or call back, I politely explain my reasons for writing the story. There's not much more to be done—if you write a lot of stories, eventually there will be someone who doesn't like one.

So that's the life of a newspaper science reporter—never a dull moment! I can't say firsthand what any other science writing job is like day to day, because I've only worked at newspapers. But from what I can guess, the general work pace is about the same for someone working at a public information office or a magazine. Each job has its own peculiarities though, and schedules will vary accordingly.

CAN I MAKE MONEY DOING THIS?

Yes! How much you make depends on where you work. I'll give some general salary ranges; these are estimates based on job advertisements and information from friends.

Newspapers: Generally the bigger the paper, the higher the salary. Small papers (circulation from 25,000 to 40,000) might pay $25,000; experienced reporters at the largest papers might make $70,000 to $80,000 or more in certain cities. Everyone else falls somewhere in between. (Note: small newspapers generally don't have the resources to devote a reporter exclusively to science, so you'll probably find yourself covering other beats as well.)

Public Information: Starting salaries can be from $30,000 to $40,000. With more experience and a promotion or two, you might earn $60,000 or more. Large, private institutions or companies are likely to pay more than public institutions.

Magazines: Depending on your experience and ability, starting salaries might be in the $30,000s. Big-time magazine writers probably make in the $70,000 to $80,000 range.

Freelancers: Fees will depend on the publication. Most freelancers are paid by the word—with word count assigned by the editor. Fees can range anywhere from $0.25 to $2.00 a word, and will also vary depending on the publication and the writer's experience. Freelancing is hard work—to earn a steady income, writers need to have several stories going at once and they need to constantly generate new ideas.

Other jobs: Salaries will vary according to institution and the writer's experience. But in general, science writers can certainly expect to earn at least as much as academic scientists, and often more.

How Do I Find a Job?

I see a lot of job advertisements for science writers, so there definitely are jobs to be had. How easy it is to get one obviously depends on your writing and reporting ability, as well as your own demands. There are more jobs in the cities, and if you are committed to a particular region, your options will be more limited. Job advertisements are posted through the standard institutional channels, and members of the National Association of Science Writers can check their Web site for openings. More about NASW later.

What Are My Opportunities for Advancement?

Once you have a job as a science writer, there are several courses your career can take. As you gain experience as a writer, you should be able to get more challenging assignments and to receive promotions and raises. Some writers transfer to more prominent publications when they feel it's time to move on. Others become editors. Lastly, if you find you really like writing and don't feel compelled to write about science exclusively, you'll have a lot of options. So many people have phobias about writing that they'll gladly hire someone who can do it without too much effort.

WILL I BE ANY GOOD AT THIS?

First of all, you have to like to write. But don't get me wrong—I don't think you need a burning passion to express yourself through the written word. However, you should enjoy the challenge, and have a natural sense of what reads well. Even if you like to write but sometimes struggle or find it hard to get started, don't let that stop you. The more practice you get, the easier it becomes. My writing speed has increased at least by a factor of 10 since I started.

Other Important Qualities:

Ego: It shouldn't be large. You can't be afraid to ask stupid questions, and you can't afford to get offended when scientists talk down to you. Believe me, some will. Also, be prepared to have your work criticized by editors. The less personally you take their advice, the more you'll learn.

Self-motivation: This should be high. Every one needs encouragement, but you should be able to work independently and take initiative. This isn't usually a problem for someone who has made it through a Ph.D. program.

Interpersonal: You should be able to work well with others. It's important to stay on good terms with your editor. And getting good information from sources requires that you can put them at ease and earn their trust. Don't worry if you've never interviewed anyone before. There are a few tricks that make it a lot easier, and with experience you'll soon become a pro.

Accuracy: Also a must. You don't want to be wrong in print. You need to have an appreciation for detail and subtlety. And don't misspell any-one's name!

An ability to work under pressure: This will come up in almost every writing job I can imagine. Fortunately, writing on deadline also gets easier with practice.

An open mind: This is a given. Unless you're writing fiction or an opin-ion piece, you have to report what your sources tell you, and it may not be what you expect. You can't make stuff up or let preconceived notions get in your way. You should be able to ask lots of questions and listen, listen, listen!

So Is This Really Better Than the Lab?

For me, without question writing is much better than the lab. Becoming a science writer has been a life-saving career move. The only thing I miss about doing research is working with my hands, and that void is easily filled through hobbies. My attention span is too short to be a scientist, so journalism suits me perfectly.

But my new career did take some getting used to. I fill out a time card now and need to be in by a certain time every day. I only get 2 weeks of vacation a year. I can't wear shorts and a T-shirt to work, or take off in the middle of the afternoon for a jog and then return to work all sweaty. My daily stress is a little higher.

But the positives definitely outweigh the negatives. I don't work weekends or evenings anymore. Also, writing is a field where you can see yourself progressing. Each type of story—short or long, hard news or feature—has its own limitations and possibilities. That means there are elements in each that you can isolate and improve on. Since leaving the lab, progress has become more obvious and tangible for me. And short of the newspaper folding—a very unlikely possibility in my case—I feel that my job security is good.

Last, I've found that my scientific background is more appreciated now than it was when I worked in the lab. If that sounds cheap (I'm supposed to be the newspaper's expert in life sciences anyway) then so be it. Having people's respect builds my confidence and makes me feel good about my job.

Where Can I Get More Information?

Check out the National Association of Science Writer's web page at http://www.nasw.org. This site will give you more information on the field. (Some links are accessible only to members.) From this site you can order a book called *A Field Guide for Science Writers*. It's written by prominent science writers and contains more detailed information than I've given here. The back of the book has a list of resources useful to working writers, as well as addresses of university science writing courses programs. Also check out the link on science writing under "New Niches" at http://www.aaas.org.

SOME INSPIRATION

If you're reading this book, you're probably already considering leaving the ivory tower. But if you are still having doubts or have a vague sense of guilt, believe me when I say it's okay to leave. When I was considering getting out of basic research, I worried far too much about what other people thought. I wondered if my professors and labmates thought I had copped out. One professor even tried to reassure me that I really could succeed in research, as if he thought I was taking on a less demanding profession. Even I wondered if I was.

But after making a living as a science writer, I can safely say I did not move to an easier job. Just because science writers write in simple language doesn't mean the job is simple. If you choose to become a science writer, you can make your career as challenging as you want.

SNAKES AND LADDERS:

The Game of Publishing

Anita Crafts-Lighty, Ph.D.
General Manager, Biotechnology Publications, PJB Publications Ltd.

"Snakes and Ladders" is an English board game in which players progress across the board according to a dice roll, up ladders and down "snakes," the winner being the first to arrive at the other side. It is characterized by sudden turns of fortune and it seems an apt metaphor for the twists and turns that the career of a scientist can take when working outside the lab. Every game has rules and, as I review how I came to be where I am and what it is like to work in publishing, I will suggest a few "rules of life" that may serve to guide those who wish to find a new career in which to utilize their scientific training beyond research.

RULE 1. FOLLOW THE DREAM
AS LONG AS IT LASTS

I decided I wanted to be a scientist when I was about 15, attending high school in Fullerton, California. It was all due to my geometry teacher, who was so inspiring that I was persuaded to remain in the science math track that I considered abandoning after a thoroughly uninspiring experience with math in junior high school. My family is very artistic (both parents were art teachers and painters and my brother became an opera singer) and I almost became a ballet choreographer until a serious bout of pneumonia at age 17 knocked me out of dancing for a few months and I discovered I didn't really miss it and would much rather do science!

During high school in the late 1960s, I became very excited about the emerging field of molecular biology and read the first edition of *The Molecular Biology of the Gene* from cover to cover like a novel. Fired with enthusiasm, I was determined to get into research as soon as possible and I was fortunate enough to be able to persuade a professor at the local state college to let me work (free) as a junior technician there to get an idea of what lab life was really like. I also enrolled for some university classes in microbiology and virology, which I found fascinating. This early experience reinforced my love of the subject field and helped me to gain acceptance to California Institute of Technology (CalTech) to major in biology in 1971.

In college, I learned three things that have really made a difference: how to work very hard, how to learn things I wasn't interested in, and how to schedule play time to balance the workload. CalTech is a small private science school, notorious as a "hothouse" environment for students. However, the intellectual stimulation of working alongside both peers and teachers who are all fascinated by science is unparalleled and it certainly taught me how to schedule my time, get through work to meet deadlines, and still have some fun.

The honor system, whereby virtually all tests are open-book and taken in your own room, was also a good preparation for later life and helped to develop a strong sense of personal and professional integrity in everyone who went there. I look back on it as a hard time (particularly physical chemistry, which I hated but managed to pass . . .) but I do not regret the opportunity to follow my dream of becoming a scientist and to work with and learn from some of the world's finest biological scientists of the early 1970s.

However, my research and teaching experiences at CalTech did make me re-examine my own talents in the scientific sphere. Working in the lab of Max Delbruck, a Nobel Laureate, was particularly illuminating. I watched with awe the insight he could bring to any scientific endeavor. I found I was able to absorb large amounts of information and organize it, which helped me to do very well in classwork, but I observed that many of my peers (and all the faculty members) were much better at seeing the core of a scientific problem and determining just what the next key experiment should be.

I may have lacked the necessary level of creativity for a really top-flight research career, but at that point I still wanted to try it. Indeed, I had little awareness then that there was any option for a trained scientist other than research; the only choice seemed to be whether to do it in an academic or a corporate environment. When I graduated in 1974, the biotechnology industry hadn't really begun and most molecular biologists thought only of a university environment. My father was a college professor and my mother was also a teacher, so academia was really the only direction I saw in which to go.

Among the choices open to me for grad school was a Churchill Foundation scholarship to study at Cambridge University in England. I had been very impressed by the English students I had met when I was a President's Australian Science Scholar. And so, in 1974 I went to England, not realizing at the time that I would not return to work in the United States for more than 20 years, if ever.

I worked in the lab of David Ellar in the biochemistry department on the structure of bacterial spore coats. In addition to doing my research (and having a lot of fun rowing for the university and my college), I gained insight into the process of obtaining grant funding and the academic rivalries that exist throughout the international academic world. With hindsight, I could see that similarly political environments had existed in all the other universities where I had worked, which somewhat tarnished my idealistic view of scientific research as the pure-minded pursuit of scientific achievement. I came to realize, as does any scientist, that one cannot separate scientific endeavor from the pressures of career advancement.

My research was interesting but of little commercial relevance and the biotechnology industry was completely unheard of in Britain at that time, but I still expected to pursue a scientific career after my second degree. However, by the time I finished the Ph.D., I realized that I had actually been working in labs for 8 years and I had had enough! I really enjoyed writing up my 400-plus-page thesis, which included an extensive literature survey, and once again I realized how much it was the desk research for information gathering that I enjoyed, not testing scientific hypotheses in the lab. My dream was changing . . . .

For a variety of personal reasons, I wanted to remain in the United Kingdom after getting my Ph.D. Since it was very unlikely that I could get a permanent academic position or funding from a British source as an American, I began to seek a job in industry. I expected, as a newly minted Ph.D., to find a research job in the food or pharmaceutical industry, but I was surprised to find that when interviewed, I was always being considered for management roles. I guess my interviewers knew more about my aptitudes than I did!

I eventually accepted a position as Senior Scientist in the technology transfer department of RHM Research Ltd., the research center of a major food and agricultural company, and I have never touched a test tube or pipette again.

The dream of becoming an academic scientist had passed, but I felt I had evaluated it carefully as an option and worked long enough in labs to know it was not the best career for me. It is important to realize that finding a career that you like and work that you are good at and enjoy is more important than sticking to any childish dream or parental expectation. Changing careers is not failure; it often requires more courage than continuing in the expected direction and can lead to much greater personal development.

RULE 2: FIND OUT WHAT YOU ARE GOOD AT

My job at RHM was not typical technology transfer work as we now understand it (in- and out-licensing of patented technology). Rather, it was a grab bag of desk research projects that did not fit into any scientific department, some of which involved collaboration with other companies. I coauthored some papers on energy analysis of the bread-making process (which showed that making bread in factories was more energy efficient than baking at home) and I worked on a project to produce a novel sweetener, as well as maintaining a watching brief on waste treatment technologies. It was all a far cry from microbiology. But I found my learning skills were useful and I got to use the information department a great deal, which led to my next major career shift.

After a couple of years of project work and management, I was eager to obtain some experience in managing people and felt restless without any real focus of expertise. I was put in charge of the Library Information and Services department, which was in need of new management. It was a baptism by fire.

I inherited some severe personnel problems, had staff reporting to me who were more than 20 years older than me and who had worked for the company for more than 10 years and were three grades senior to me due to an interesting matrix management system, plus of course, I had no formal training in information science or librarianship! I was, however, a longtime library enthusiast and had classified all my books in the Dewy decimal system at age 10 (complete with shelf marks on the spines) so maybe I should have realized that I had libraries and books in the blood.

I enjoyed bringing the card catalog up to date and learning how to do online literature searches, and I introduced a records management and archive system for project files and laboratory notebooks. However, after I straightened out as much as I could in the department, the expected promotion was not forthcoming. I soon decamped to Britain's first biotechnology company start-up, Celltech Ltd., feeling that the science there would be much more interesting and the longterm career prospects better. Little did I know at the time that this would lead to yet another career shift within 5 years.

RULE 3: MAKE YOUR OWN OPPORTUNITIES

I pretty much created my job at Celltech. I contacted the founders who were setting up the company and persuaded them they needed an information service. Once I was firmly ensconced there, I generally set my own budgets and goals. It was rather daunting to walk into an empty room and realize that you had to build a complete service from nothing. But it was great fun and, since Celltech was a small start-up company, I was very close to the users both in R&D and in marketing and management. I learned a great deal about how to run a business and how to motivate people by observing the senior management team, and I was able to get involved with many other aspects of the company, such as research project planning, computing policy, public relations, and records management. I was also fortunate in some ways not to have an established library, because I could computerize everything from the start without having to worry about card catalog conversion.

One of the things I soon noticed was that whereas the scientists were well-served by various information providers, the business-oriented databases of that time had very poor coverage of the biotechnology industry. They did not abstract any of the numerous newsletters that had been started to serve this emerging sector. We began an internal service to summarize these newsletters for Celltech's managers because we found that the content overlapped to a considerable extent and it was very time-consuming for management to try to read them all.

Our news abstracts were circulated to external scientific advisory board members and before long, biotech industry competitors like Genentech and Biogen in the United States were calling us to find out more about our service. We felt we had a product opportunity on our hands and, although the company had had no particular intention to diversify into publishing, any source of revenue was helpful in the early days of rapid cash burn. We decided to launch a printed product called *Abstracts in BioCommerce (ABC)*, which would provide a comprehensive monitor of events in the biotechnology business. *ABC* began publication in August 1982.

We set up a joint venture with IRL Press Ltd., a publishing company that had marketing experience in the sector and prior experience of publishing scientific abstracts, as well as printing facilities and administrative experience in circulation management. Although the scientific abstracting side of IRL's business had just been sold to Cambridge Scientific Abstracts, IRL was interested in moving into business information. The cooperation worked well for the first 2 years. In retrospect, I learned a great deal about the publishing industry and marketing techniques in that period, and about the importance of meeting production deadlines.

However, by 1984 it was clear that abstracts must be available in electronic form for archival value, in addition to the print format. IRL Press was not interested in exploiting that side of the business and Celltech decided to go it alone.

The electronic database version of *ABC* went online with Data-Star in December 1984, and I found myself increasingly called upon to market the online product. The abstracting activity, while profitable, was a significant resource load on the information department at a time when Celltech was beginning to focus its activities more closely in certain therapeutic areas and needed to reduce overhead cost.

I was also feeling a little trapped in my information department role, with no immediate career prospects and a growing interest in learning how to run a company. In 1985, I proposed to Celltech that I acquire the rights to *ABC* and an associated company database developed for indexing purposes, and I also proposed a buyout of IRL Press's rights to the title to make it easier to develop the print and online versions concurrently. Once again, I had to make my own opportunity and face the empty office to create something from nothing.

RULE 4: BELIEVE IN YOURSELF

Starting up in business is never easy and I really put the lessons of college to the test. I was fortunate in that I was not really starting from scratch—I already had an established customer base and revenue stream. However, I did have to establish independent production systems, employ editorial and administrative staff, find and equip premises, and put in place financial control systems. I still remember drinking champagne at 11:30 P.M. on the first production day, after we had finally made the printer work!

Learning double-entry bookkeeping to keep the accounts was far easier than physical chemistry, and was essential to understanding how the business was going. A formal business plan was not necessary in my case, since I was not raising any external capital. In any case, you must have at least a mental plan and a simple cash flow budget. Prudence in expenditure is wise; you cannot expect your own fledgling business to deliver the same remuneration package as your previous salaried job.

Again, I was fortunate to have one major customer in the financial sector who was willing to negotiate a 4-year data supply contract payable in advance, which funded the capital investment I needed to set up the very expensive computer systems needed for electronic publishing.

I found the first few months of running my new company, BioCommerce Data very hard work (not helped by the fact I was persuaded to write the second edition of a book, *Information Sources in Biotechnology,* for another publisher!). But gradually we settled into a routine and were able to expand both staff and premises. In 1988, we began producing the *U.K. Biotechnology Handbook* from a larger worldwide database, in collaboration with the Association for the Advancement of British Biotechnology (this later became the BioIndustry Association). We also began publishing another periodical, *Biotech Knowledge Sources,* in collaboration with the British Library. In 1995, we introduced a sponsored bulletin, *BioCommerce Financial Abstracts,* in collaboration with Ernst & Young.

While I had always been very dedicated to my job and tended to work very long hours for my employers, the demands of running your own business are never-ending. One must constantly prioritize the work to make sure the important things are done—not just the urgent ones—and to leave enough personal free time for relaxation to ensure that you can function efficiently. Personally, I found it very valuable to have a separate office. If I worked from home, I would never stop working, although I do still bring work home!

Running a small business means looking after everything. Of course you can, and usually must, employ some professional advisors (accountant, lawyer, etc.) but they will never understand your business as well as you do. To get the best from their expertise, you need to interact closely with them. Make sure you read and understand every contract. Make sure you know what's behind every number in your audited accounts. You are the one who will have to explain them to a bank manager or future purchaser. And you are the one who will have to deal with the consequences if there are problems. You must have confidence that you can learn enough to handle it all and you will constantly have to solve new problems. The buck always stops with you. Just because you don't know how to solve a problem when it arises, doesn't mean it you can't solve it.

This is especially true of support systems such as computers. We are all increasingly dependent on computer systems. These days, publishing is almost totally computerized. Don't even think about it as a career if you are not completely computer literate and reasonably comfortable with the rapid pace of technological development in this area.

In publishing, computers are critical to production, as well as useful in accounting, corresponding, financial planning, marketing, and so on. Make sure you know how to install and use your software, and that you have access to both hardware and software support from reliable external· sources so that when the problems do exceed your expertise (and they are bound to at some point) you have someone to turn to. Learn to read manuals and ask logical questions.

Probably the most important thing I learned was ALWAYS BACK UP YOUR COMPUTERS! More than 50% of businesses fail after a major computer disaster. In 12 years of business, I have experienced theft, flood, unexpected hardware failures, and user and software errors that time and time again required the use of backup tapes. Never have just one tape, but keep a series (a common system is daily tapes for 1 or 2 weeks and monthly archives) and never rely on just copying files from machine to another in the same office. That's fine in the case of accidental file deletion or machine failure, but no use if the whole office burns down or if all the machines are stolen. Make sure your backups are stored at a remote site, preferably in a fireproof safe. In database publishing, your data is your business; lose that and you have no business.

However up-to-date you are now, expect to cope constantly with change in computer systems. In the last 12 years, I have had to learn five word processing systems, six computer operating systems, and I currently use three different Internet providers and e-mail systems. That doesn't include the major changes that new versions of software introduce and other packages for presentation graphics, typesetting, backups, and spreadsheets. You must budget the time and money to keep up with technology where there is a business justification to do so. Change for the sake of change is never worth it, but there is much to be said for always migrating to the latest software version and operating system as soon as it is clearly stable and popular.

THE PUBLISHING BUSINESS

Publishing is essentially the dissemination of information. There are broadly two types of publishing: primary and secondary. Primary publishing is the publication of original articles, newsletters, magazines or books, scientific journals, and textbooks. Generally, most of the authors and editors of scientific publications are employed outside the publishing company, in academia or in companies. They are paid a royalty on sales or a fee for work done (such as expert reviews or contributed articles to books), or in some cases a retainer (for example, to serve on the editorial board of a journal). Scientific journal articles are generally contributed without any remuneration, as part of the professional process of research, although review articles may be commissioned.

Magazine and newsletters publishers usually employ in-house editorial staff to write and compile their publications, but typically also use other journalists on a freelance basis. Book publishers employ commissioning editors in-house to develop new titles and journal editors to oversee those publications.

The publisher absorbs all the costs of getting the information to the readers (typesetting, printing, distribution, and marketing) and generally takes most of the profits. When print publishing began, this was a technologically complex and expensive business involving high capital investment for skilled typesetters, large printing presses, physical distribution systems, and so on. Today, with the advent of the Internet, electronic publishing is easy and cheap, and the role of primary publishers, many feel, is becoming one of the official archive, a repository system whereby your publication is guaranteed a permanence and peer review.

However, this is a time of flux for publishing and the true impact of electronic primary publishing in the sciences is yet to be fully realized. It is probably an interesting time to get into the field but perhaps one of economic contraction. The situation is perhaps a little less acute in business information, but newsletters and newspapers must now compete with free news services providing the press releases that were once the basis for their articles directly to readers, so they too are seeking ways to add value to their offerings. The key probably lies with the original content and analysis.

Secondary publishing is basically the process of indexing, abstracting, and organizing primary publications, such as what we did at BioCommerce Data. Secondary publishing is also in a time of flux as new technology makes searching full text databases easier, quicker, and more useful. However, the exponentially increasing amount of published information makes the need for filtering and summarizing systems ever more important and so there is likely to be a continuing role for some such services.

Publishing directories, another activity of BioCommerce Data, is a bit of a hybrid type of publishing. Most contain original information but they involve the collection and organization of large quantities of data, and while some directories are compiled by external authors, most are prepared in-house by the publishers. Abstract databases may use all internal staff or some freelance or home-based workers, often graduate students. Abstracting is often an entry level job in STM publishing for people with an undergraduate or a higher degree in science and no desire to continue in the lab.

Most scientific, technical and medical (STM) publishing is based on subscriptions. Some magazines and directories in the sector are distributed free, with their costs paid by advertisers, exactly the same concept as the free local newspaper but with rather different content.

A publication may be of excellent quality and content, but if it cannot be published profitably, no commercial publisher will continue it. This is a big difference from the mentality of basic research and you need to be comfortable with the need to market a product as an essential part of the publishing activity.

WHAT ARE THE JOBS WITHIN PUBLISHING?

The tasks of publishing can be broken down into five broad categories: market research, product development, production, sales and marketing, and competitor monitoring. These tasks describe a circle linked to a product life cycle. First you identify whether a product is needed, then you have to create it, then you have to make it and successfully sell it, and finally, you have to change it in response to competition and to respond to changing market needs, which brings you back to research again.

Market research can be conducted formally through interviews and questionnaires or more casually through an ongoing process of dialogue with your customers and target market sector. However you approach it, market research will seek to establish: Is there a need for this type of information in this form? How much will readers pay for it? What content and form of presentation do they prefer? Are competitors already providing something similar? and, if so, What are the limitations or disadvantages of their products in the eyes of their customers? What would give your product a real advantage in the marketplace? It is very important in any research aimed at revising existing products to talk both to your current customers and to nonpurchasers.

Product development is a mixture of people and content. You need to create the right collection of information by, for example, commissioning articles on hot topics for a book (like the editor who commissioned this book, for example), covering all the right organizations in a directory, attracting the right sort of authors to a journal, or developing a really comprehensive database.

To do this, you need a good editorial and design staff because presentation is also important. You may need to design a snazzy Web site, you may need to have state-of-the-art retrieval software or just a nice-looking print publication. The style needs to be appropriate to the readership— academic books, business newsletters, and advertising-based services all take different approaches depending on the expectations of the typical reader. Even within these genres, publications vary greatly in style, with some being very factual and others deliberately putting forward thought-provoking opinions or promoting the activities of advertisers and sponsors.

Whatever you are presenting, you need a good editorial staff composed of people who understand the subject matter and the marketplace and who can deliver quality work to deadlines. Finding these people can be quite difficult, and all publishers are always on the lookout for new staff who wish to move away from a laboratory career and have the right sort of personal qualities. To keep and motivate the editorial staff is a managerial challenge that requires a constant balance between the knowledge base and the operational continuity that is achieved by having the same long-time editor versus the fresh approach that a new editor can bring.

Once you have developed the publication, you must keep it rolling out on time, and getting either very creative or very detail-oriented editors to do this can also be challenging. Written operating procedures for routine tasks are useful in the event of staff changes, but the key factor here is planning and scheduling, plus an attitude that the product *must* come out on time! It is easier sometimes to keep to such schedules with in-house staff. Editors who use busy external authors must develop high levels of diplomacy and persuasiveness to obtain their manuscripts.

Now that you have a regular publication going, and often before the product actually exists, you must market and sell it. The distinction between marketing and sales is blurred. Essentially, marketing is broadcast publicity, promoting the product usually by direct mail and on-page or loose insert advertising in other print publications aimed at the same sector. Pages on the World Wide Web are also a part of the mix, as is a presence at relevant exhibitions and conferences. Marketing also encompasses preparing sales materials, such as flyers or leaflets, promotional letters, exhibition panels, sample or demonstration copies, Web pages, and so on.

Sales is the actual process of making the sale, which is usually a one-to-one transaction. This may occur in person, by mail, e-mail, fax, telephone, or via a third party such as the distributor of an online database or a foreign agent. Direct mail campaigns are the principal marketing method for print publications, and looking for sources of good lists of potential subscribers is a key marketing activity. These lists may come from other publishers, conference delegate lists, purchasers of other titles from the same publisher, industry and professional associations, and other sources.

Sales will only involve personal visits for large-value items, such as major corporate database subscriptions or large advertising contracts. With the exception of exhibitions, most direct sales activity in publishing is telesales, and even that is mostly for advertising. It is simply not cost-effective to visit purchasers to sell a $50 book or even a $500 newsletter or journal, so good marketing is essential, with much of the sales activity really being just order taking.

In developing a publication, you must allow for a launch phase during which marketing costs are extra high. Once a target sales level is reached that guarantees profitability at an acceptable level (taking into account all staff, production, distribution, and ongoing marketing costs), a continuing marketing activity must be maintained to compensate for business that will be lost due to personnel changes, changes in corporate interests, budget cuts, and so on. The ongoing efforts usually aim to slowly increase the subscription or user base over the years and to replace losses of from 15 to 30% per year.

Once your publication is established, you must constantly monitor competitors. If you have a good idea and if you are the first to develop it, you can be sure that someone else will copy it, and probably improve on what you have done and/or undercut you on price. They may or may not have a better marketing approach, but they can learn from your mistakes and they no longer have to introduce the product concept to the marketplace.

At present, a new type of competition in publishing is the increasing amount of free information available from the Internet. In response to such pressure, it is vital to ensure the highest levels of quality and value-added as well as easy access to your information. You cannot expect to not have competitors and you must not be frightened by their existence into giving up, even if they seem to have better resources than you do.

Over the years at BioCommerce Data, we have outlasted or superseded a number of competitive products that initially seemed to present serious threats. Never be complacent, but don't panic either! Decide how you can have a better product that provides value for the price, market it equally effectively, and you should be able to retain an adequate market share.

RULE 5: KNOW YOUR LIMITS

If you want to get into publishing, you need to consider carefully what role you would prefer. Do you seek the jack-of-all trades life of a small independent publisher? Do you have the patience and attention to detail necessary to be an editor, abstractor, or indexer? Is your forté the conceptualization of a publication and author selection (the role of a commissioning editor)? Are you attracted to the regular production and manufacturing aspects of publishing, or perhaps to the sales and marketing side? Do you prefer print or electronic products? How relevant is your background? For example, do you have specialist expertise in a particular field that you could exploit to develop a new database or journal, or to write a definitive reference work?

Think carefully about what type of person you are. Generally outgoing, gregarious, and creative individuals are better suited to jobs as journalists or in sales, while more analytical, organized, and introspective types tend to make good editors. A good test is to consider whether you are more of a "people person" or a "facts person" (which would you rather work with, given the choice). However, there are exceptions to every generalization and much will depend upon the environment in which you find yourself.

In 1996, after 11 years of independent, operation, I sold BioCommerce Data to PJB Publications Ltd., a much larger U.K.-headquartered business information provider, known worldwide for its newsletter for the pharmaceutical industry, *Scrip,* and database of drugs in development, *Pharmaprojects.* PJB also publishes other newsletters, numerous reports, and directories for the medical devices and diagnostics, crop protection, and animal health markets. So, why did I sell and join a larger organization again?

The reasons were numerous but they serve to illustrate many of the issues of publishing. First of all, as an entrepreneur, I needed an exit route for the investment I had made in my own company. I had been lucky in that I had not had to raise external capital (and exchange that for an equity share in the business), so I had no investor shareholders to consider and no debt finance (e.g., a bank loan) to pay off.

However, I had deliberately retained virtually all the company's profits to help its cash flow and facilitate new investments, and had not taken a full market-value salary from the operation. After 10 years, I felt it was time to cash in this investment. I had built the business up to a point where it was not wholly dependent on my editorial input on a day-to-day basis, and it had annual revenues sufficient to interest a larger company able to fund further growth potential. These were very important factors in finding the right buyer.

Selling a company properly is a fairly time-consuming and expensive business. My legal costs were 10% of the sale price, their costs including "due diligence" work to review the company's financial position and legal agreements were even more, and the process took about 6 months to complete after the initial discussions. If you are too small, no one will bother with this, although they might acquire a title (i.e., one book or journal), which is an approach generally involving less risk than buying the entire business. If you seek to do the latter, you must find an appropriate partner. We were approached by several more scientific database publishers over a period of some years, but ultimately we felt that being part of a business information-oriented group gave us better access to our markets.

Second, I felt personally trapped, unable to expand the business further from my own resources of capital and time, and finding it difficult to attract good staff to a small operation in times of economic uncertainty. At the same time, I was unwilling to give up an activity that still had growth potential and that was relied on by thousands of users. We were going through a slightly difficult patch with some changes in clients and personnel, and would have found it hard to expand when we needed it most. I found that I wanted to be part of a larger organization again, where I could call on the support services of other departments, particularly for marketing and sales activities where critical mass is so important.

To grow, we needed in-house advertising channels with high circulation, we needed sales staff for both the U.S. and European markets who could call on large clients to negotiate customized data delivery deals, and we needed more expertise in large-scale direct mail marketing. In addition, we hoped that administrative services like personnel, accounts, purchasing, and so on, would be more cost-effective as part of a larger enterprise.

Finally, the publishing industry is undergoing a great deal of consolidation with many observers claiming that by the year 2000, there will be only five or six publishers of note. I felt it was unlikely that BioCommerce Data would be able to continue to compete successfully in this environment, especially as some of our competitors were already being acquired by larger organizations.

After so long on my own, I felt ready to go back to an environment where I could share and discuss the day-to-day problems of management with peers and have other people to bounce ideas off of for strategic development. I had never really found a business partner who could share the load fully and whose skills would complement mine. I had difficulty balancing the needs for sales development with editorial responsibilities and all the day-to-day administration needs of a small business.

So how has it worked out? When I sold BCD, I agreed to continue to manage it as a wholly owned subsidiary of PJB, while also taking on the role of General Manager, Biotechnology Publications for the parent company. In this capacity, I am charged with developing and consolidating our publishing activities in the biotech sector, and I have budget responsibility for another biotech publishing subsidiary PJB recently acquired, BioVenture Publishing Inc. in California.

A year into this new phase, I don't regret the decision to sell but I'm working harder than at any time in my life since I started my own company! I am enjoying the new challenges, but inevitably some things have been easier to adjust to than expected, others harder. In any large company, some individuals are more capable and more dedicated than others and it takes a while to learn what people are like, or even who does what. I found that being physically integrated with the rest of the company helped a great deal in this respect.

Many scientists who are considering a move into publishing ask me what it is really like in comparison to the lab. The publishing environment is probably more similar to corporate research than to academia. It's more than a full-time job for most people. Positions are salaried, not paid by the hour, and most staff are expected to work significant amounts of unpaid overtime—the more senior you are, generally the more extra time is usually put in.

You are expected to take your responsibilities very seriously—you need to be available for contact even on vacation, to work weekends if necessary to meet a deadline, and so forth. This may sound pretty familiar to researchers but the hours are generally less flexible—no long summer breaks from teaching, and an expectation that you are always in the office at least from 9 A.M. until 5 P.M., even if you are also working until 8 P.M.! The office environment is somewhat more formal in terms of dress, and of course the work is mainly done sitting down! Meetings are an unavoidable time sink and, in a larger organization, memos and other internal correspondence (weekly or monthly reports, etc.) are also a perennial feature.

At the editorial level, the work may be very routine but one must remain very alert to detail while still delivering quality work, and must ensure that all production deadlines are met. As a manager, work is generally somewhat less focused, with lots of minor actions to track alongside the more strategic work and longer term projects, which gives one a contant sense of juggling (too) many balls at once.

Constant rescheduling of activities is necessary to keep things prioritized appropriately, and it can sometimes be difficult to find the time to do large projects like business plans or annual budgets. One can expect a certain amount of administrative support with things like purchasing, travel arrangements, and filing, but many managers today handle all their own correspondence by e-mail or directly rather than by using a secretary. I spend a lot of time interacting with other departments, such as personnel, accounts, creative (design) services, and so on, as well as managing my own staff. Because my position is at a fairly senior level in the company—I report directly to the Managing Director, who is also one of the two owners of the company—I have a considerable degree of autonomy, but obviously I am now accountable to others for my actions in a way that I was not while running my own business.

Management (of anything, probably, but certainly in publishing) seems to divide into four categories: the mundane but necessary details (authorizing payments, checking that actions have been carried out by others, claiming travel expenses, etc.), the emergency fire-fighting (organizing temporary staff when someone quits or is sick, getting the computer systems running again, and so forth), meeting defined objectives (hitting sales targets, carrying out preagreed projects or tasks, publishing on time, etc.) and planning for the future (which can be everything from doing next year's budgets, to new product development, getting competitive intelligence, or sniffing out potential acquisitions).

Balancing these is the trick. You must be organized and while enthusiastic and optimistic, you must also be realistic about time scales and the possibilities. You must be pragmatic about the support available from others, whether it is your own staff or colleagues in other departments—work together when you can motivate everyone to achieve shared corporate goals but find other solutions when you must, which sometimes means to do it yourself! Be wary of the temptation to abdicate responsibilities to others if they are not delivering for you. The buck still stops with you as a manager but you can't do it all personally. Balance and determination are the key. I rarely take "no" for an answer and I will not be defeated by machines, however recalitrant a software package may seem!

RULE 7: COMMITMENT LEADS TO ENJOYMENT

I cannot enjoy something I do half-heartedly. I believe that to be happy and motivated in what you do, you must be committed to do the best you can at all times. Whatever the frustrations of the day, I get satisfaction from solving the problems, and from moving projects forward. I don't do much that is especially creative in the way that developing a new hypothesis for a biological process is, but I get a lot of satisfaction from seeing each database update completed, each new issue or edition published, each new piece of marketing literature created and used, and each new title launched. I enjoy collecting and organizing information and developing systems to turn data into information and facts into knowledge.

If you share a similar outlook, you may well find your career in publishing—but while you climb that ladder, watch out for the snakes!

5

· · · · · · · · · ·

BROADCAST SCIENCE JOURNALISM:

Working in Television, Cable, Radio, or Electronic Media

· · · · · · · · · · · · · · · ·

Eliene Augenbraun, D.O., Ph.D.
Chief Executive Officer, *ScienCentral;* Co-Executive Producer, *Science Friday* Website

Karin Vergoth, MA
Co-producer, *Science Friday*

Did you ever dream of hearing your voice on the radio talking about science or hosting your own TV science show? But wait, you may say, there are only a handful of science voices out there! Does that mean that broadcast is no place for an aspiring science journalist? By no means!

There are lots of ways to work in broadcast or electronic media. National Public Radio's *Science Friday,* for example, has two producers (one of whom is Karin Vergoth). They do the research, preinterviews, and writing that make the show a success week after week. And *Science Friday* also has a web site (www.sciencefriday.com), with supporting material assembled and written by a web producer every week. (Eliene Augenbraun is co-executive producer of the *Science Friday* web site with Ira Flatow, the host and executive producer.) And don't forget about all those science producers, researchers, and writers on television, commercial radio, and the Internet.

Nearly every broadcast show has an executive producer who gets the money together and manages the projects, producers who do most of the research and writing, talent who appear before the microphone or camera, plus technicians and advisors. News shows typically add editors, and large organizations have business managers and fundraisers. There are plenty of ways to bring your science skills to a broadcast job.

One of the first things you need to figure out is what kind of work—especially what kind of deadlines—you feel comfortable with, because when the show must go on, it won't wait for you to be ready! If you're a perfectionist, you may hate daily deadlines. Monthly deadlines may be hard for procrastinators. Do you like talking? Writing? What frustrates you? Some days you may not want to talk to anyone, but talking to people is what a producer does for a living.

There are several kinds of jobs in broadcast science journalism, and if you look closely, you may find one that's a perfect match for your skills and interests. Here's a rundown on some of those jobs and the activities they entail. Because there is such a broad range of jobs, and because salary varies with the details of the job and the specific employer (public radio vs. for-profit TV, for example), we have not included information on salaries. Once you have identified a position that interests you, contact others with similar jobs to get information on salary ranges and typical benefits.

REPORTER

Reporters bring the latest science news to viewers or listeners of TV and radio stations. As a National Public Radio (NPR) science reporter, Joe Palca works on two or three stories in a typical week, each 3½ to 5 minutes long. He covers a wide range of science and medicine topics for NPR, so in one week he may be tackling stories as diverse as the weather on Mars, the discovery of a "cell clock" gene, and the president's new AIDS announcement. For many TV and radio reporters, the bulk of the science stories they cover are consumer health stories.

Story ideas come from reading newspapers, wire service stories, and journal articles, and sometimes they come from a source's phone call. About half the stories Joe ends up doing are assigned to him, and the other half are ideas he pitched to his editors.

In an ideal world, a reporter armed with background material would find and interview sources for a particular story, then put together that piece, and move on to the next. In the real live news business, reporters need to be flexible enough to work on multiple stories at once, splitting their attention between different sources and breaking news events, and still meeting assigned deadlines.

Believe it or not, skills learned in the lab do come into play as a science reporter. Most obvious of course, are the science training and research skills, which can be invaluable for a reporter trying to understand the science behind a particular story, or exactly what a piece of research entails. Being able to grasp ideas quickly and turn them into a story just as quickly helps when it comes to meeting the daily deadlines of the news business. Joe found that his experience teaching statistics to psychology majors was good training for telling his stories to the listeners. Teaching and radio reporting require the presenter to get the listener's attention by making the information interesting.

One place where science training may get in the way, however, is when a reporter must decide which information to present. Scientific audiences demand a high degree of precision and accuracy, but reporting science to a general audience often requires sacrificing some precision to get the basic message across. Keep in mind that in radio and TV, a listener can't back up and re-listen to something that didn't quite sink in the first time.

The rewards of working as a broadcast reporter, as you might imagine, are different from those of doing research. In journalism, they're more immediate, while anyone who's worked as a lab tech knows that the rewards in science can be a long time coming.

Joe Palca, now a science correspondent for National Public Radio, explains how he moved from science to broadcast journalism:

I was a grad student at the University of California at Santa Cruz, studying human sleep physiology. I wanted to do basic brain science, but there were no research jobs to be had, and money was scarce unless you were an M.D. I had completed all the requirements for my degree except for the dissertation when I saw an ad in a magazine for the AAAS Mass Media Science and Engineering Fellows Program. I was accepted into the program and assigned to work with a science reporter at WDVM-TV, the CBS affiliate in Washington. Because of union rules, I couldn't write the stories or appear on camera. I could, however, be an associate producer, which generally entailed doing research and tagging along on the stories, newshounding, booking interviews, going out with a video crew, and shooting segments.

After my 10 weeks as a producer were up, I finished my Ph.D. and applied for a science job, but he was facing a tough job market. I decided to try for a journalism job. I joined media organizations, and applied for entry-level jobs in TV and radio, landing a gig writing and researching pieces for Health News *on KQED in San Francisco. A fellow member of the Media Alliance Group was a producer for ABC. Using that contact, I took a writing test at ABC and got hired as a news writer, where I wrote for the evening news.*

My next job was as a part-time desk assistant at the NBC station in Washington, where I did a lot of grunt work, but I also got a feel for the news business. Finally, I ended up back at WDVM-TV working as an assignment editor, sending reporters and their crews out to cover a story. A year later I quit, moved to England, and worked with a friend writing copy for a PR firm. Six months later, I was back in Washington looking for a job. There were few entry-level jobs in science writing, so I did some freelance writing and producing.

While I was working as a science producer for a general assignment reporter at WDVM-TV, where I picked stories, set up interviews, and briefed the reporter, an editor at Nature *called and offered me a job as their Washington correspondent. I wrote news stories for* Nature, *and then I moved on to* Science *magazine, spending 3 years in each place. Then in 1992 my current boss, Anne Gudenkauf, asked me to come to NPR, where I've been ever since. Currently, I am a science correspondent and a special correspondent for* Sounds Like Science *on NPR.*

PRODUCER

Science Friday is a live science talk show on National Public Radio. Every week the show presents the latest science news by bringing the scientists themselves on the air. Listeners call in to talk with the scientists and with the host, Ira Flatow. Karin Vergoth produces 1 hour of the 2-hour program every week.

What goes into producing a science talk show? In a nutshell, researching and selecting program topics, booking and pre-interviewing guests, writing the script, questions, and promotional copy, and directing the live broadcast.

Choosing a topic for Science Friday is a collaborative process between the producer and the host, and it requires staying on top of the latest advances in all fields of science. That means reading almost everything in sight, from the standard science publications such as *Nature* and *Science* to press releases, science magazines, and newspapers. Because the show is almost entirely live, there's not a lot of writing, at least compared to print journalism. Scripted parts of *Science Friday* include the "billboard" that's heard at the beginning of the show, the introduction, and the questions (which are, of course, subject to change as soon as the guests and the host come together and the show begins).

Directing a live call-in talk show has its own challenges, from making sure that the guests are where they're supposed to be at show time (usually in some remote studio) to making sure that the host and the engineer hit their cues at precise times throughout the show.

Any job has its downside, and the downside of being a producer for a live show is that you don't have final control over your product. After spending a week putting together an hour of *Science Friday*, a producer's work ends before the show, and it's up to the guests and the host to finish it off. In the end, the final product might bear little resemblance to what the producer had in mind. Of course, says former producer Karen Hopkin, nothing really compares to the rush of working on a live broadcast, shutting off the microphone and knowing that you made it through another show and that it all worked.

TV production is not a lot different, except that it is complicated by the need for images and the occasional prop. A lot of the video used in a typical news story is kept on file with the organization that made it. For example, NASA will provide images free of charge to news organizations. Other video needs to be made on location for the story that is being produced, which requires a camera and a sound crew. Obtaining permission to use video and locations for TV broadcast requires a resourceful producer. A TV producer organizes all the people, obtains images, books experts and locations, and writes the story.

How do you end up in a producer's role? Karen Hopkin, former producer and senior producer with National Public Radio says her first exposure to radio was as a science reporter and an American Association for the Advancement of Science (AAAS) mass media fellow, where she wound up at WOSU in Columbus, Ohio, in one of two radio slots. Despite an intensive 2-day orientation to science journalism, all she could remember during those first tense interviews was not to say "uh-huh" and "mm-hmm" during the taping. As Karen tells it:

> *Well, I got through my first interview okay. Then came time for me to put together the story. After sitting at my desk with a reel of tape and about six pages of handwritten notes, I cornered my editor and asked "How do I, like, actually write a radio story?" "Well, Karen," my editor began, "You'd write something like, 'In today's issue of* Whatever Journal, *Some Scientist or Other reported that such and such happened." With that input, I wrote the story. And recorded it, here and there sticking in an "actuality," the bits of tape that feature an Actual Scientist saying something pithy about such and such.*
>
> *It was a great summer and I learned a lot about radio. I put together 20-some taped pieces. I guest-hosted and sat in as a guest on four live talk-show programs. It was a great deal of fun and a lot of hard work. But I also felt that radio was so superficial. After all, I was trained as a biochemist, and in a 2-minute story I didn't really get to dig in and talk about molecules. (I did one piece on subatomic particle physics, but that was because my editor thought the words sounded really cool—muons and gluons, bosons and quarks.)*

So I went back, finished my Ph.D., and eventually found my way into print journalism. Finally, I could write about molecules. But I missed radio, because it was so dynamic, so rich, so textured, and so much fun. I missed the sounds, the voices, and the personalities. In my experience, people who do radio are pretty much characters, from the anchors to the engineers. I found my way back to radio via the NPR program Science Friday, *which I produced for 3 years. My* Science Friday *gig was part-time and during my years there I did a lot of freelance writing for print, particularly for the* Journal of NIH Research, *where I now work full-time.*

For me, moving from bench science to journalism was a good choice. After years in grad school, I realized that I didn't want to focus on a single scientific problem like a laser beam for the rest of my life. I wanted to ask lots of questions and learn everything I could about how the world works. Writing about science allows me to live vicariously—finding out about the very most interesting things, reveling in the excitement of other scientists' discoveries, and then moving on to the next thing. I enjoy science. I like telling stories. And in my experience, writing about science is the best way to combine the two.

Karin Vergoth, a producer for National Public Radio, considers herself an almost-scientist who became a science journalist:

I first became interested in science journalism when I was a doctoral student in industrial and organizational psychology at New York University. In the course of my class work and research, I began to realize that although I enjoyed the field I was studying, I didn't really want to do what the people with the Ph.D.'s were doing. I was good at doing research, and paying attention to the details, but what I enjoyed more was thinking about the results of the research, how they fit with other studies, and what that said about the way people ticked.

At the same time, I also realized that although academic researchers were learning a lot about how people work in companies, most of what they knew wasn't making it beyond academic journals or meetings. There was a gap between what university researchers and managers knew about people's work behavior and motivation, and I decided to bridge that gap by communicating the research to people outside the field.

After finishing my master's thesis, I quit the program and started working as a consultant. I also enrolled in a writing program at DePaul University in Chicago. After a 1-year stint as a research project profes-

sional at the University of Chicago, where I managed a National Institutes of Health grant, I realized that my interest in science stretched far beyond the field of psychology.

The following year, I entered New York University's Science and Environmental Reporting Program. While at NYU, I worked as an editorial intern at Psychology Today *magazine, where I wrote a dozen news stories about current psychological research (and did more than my share of making copies and answering phones). After graduating, I took a temporary position as an editor at* Scholastic, *where I wrote news and feature stories for a kid's science magazine,* Super Science Blue. *I never thought I would be a broadcast journalist, but here I am. Since May of 1996, I've been a producer for* Science Friday, *where I've been lucky enough to report on a broad range of science topics—but still not industrial psychology.*

EDITOR

In a typical newsroom, editors either assign stories to reporters or approve story ideas suggested by reporters or producers. They work with reporters to decide what approach should be taken, how the story should be written, and whether, in the end, the story will make it on the air, or be bumped for more newsworthy stories.

Richard Hudson serves as the science editor for a public TV station's many projects. In this role, he's involved in brainstorming activities for program segments, looking for clever hands-on science to offer viewers, and wrestling with the best way to present a complex scientific idea.

John Keefe, science editor at Discovery Channel Online, talks about how he got into the business:

Back in seventh grade, a civics teacher told my class that the jobs most of us would have as adults didn't yet exist. Boy, was he right. Today, as science editor for Discovery Channel Online, *I'm working in an entire medium that emerged just a few years ago. I was always interested in science, journalism, and technology, but I never could have imagined taking the path I took to get here. My career path is an example of what can happen when you let your interests and curiosity guide you through uncharted territory.*

My college years were steeped in newspaper lore, both at the University of Wisconsin-Madison's student-run daily newspaper (there was only one at the time) and for the city's morning paper. At the student paper, I tried everything imaginable, including taking pictures, writing, managing the production team, and editing. At the citywide paper, I was a part-time general assignment reporter who weaseled his way into covering cops on the weekends, too.

I also fulfilled my science appetite by taking introductory classes in a variety of different fields including astronomy, genetics, and biology. Basically, I learned just enough to get the basics, but not enough to really know anything, something that turns out to be valuable in the world of journalism.

My cop-reporting experience helped me nab a job at the Racine Journal Times, *located in a city just south of Milwaukee with one of the highest crime rates in Wisconsin. Eventually, life on the death-and-destruction beat wore thin, so I started attending science-writing conferences with the hope of covering more science. When, at one of these events, I saw a poster for a science-writing radio internship in Washington, D.C., I jumped at the chance. I headed for the American Association for the Advancement of Science, where I wrote 90-second radio features that aired on the Mutual Broadcasting System.*

Over time, my boss and I heard that a children's radio station in Minneapolis was planning to take its format nationwide. We thought, Hey, we should make a kids' science show for that network. A few years, a few grant proposals and a few million dollars from the National Science Foundation later, Kinetic City Super Crew *was being heard across the country each week and I was in charge of all of the production aspects as senior producer. (One of the perks of co-writing your own grant is that you get to nominate yourself for a top job). The* KCSC *science adventures, performed mostly by child actors, still entertain and educate each week nationwide, and this year the show won a George Foster Peabody Award, one of broadcasting's highest honors.*

My work on Kinetic City *helped me land a couple of freelance radio gigs, including two half-hour documentaries for the* Soundprint *program on National Public Radio. For one piece,* Soundprint *asked that I prepare some Internet resources for them. I did, and sent it to them in HTML, the programming language of the net, because I had been tinkering with web stuff on the side. Later,* Soundprint *landed a deal with Discovery Channel Online (www.discovery.com) to produce a live Internet talk show each week on the Internet. They remembered my web experience, needed a pro-*

*ducer, and asked if I'd be interested. I did it as a freelancer, keeping my
day job at* Kinetic City.

*About 6 months into the project, Discovery said they were creating a
new science editor position and asked if I would be interested. I moved
there full-time. Since then, I've produced/edited dozens of science stories
and several major events, all tailored for the Internet and its interac-
tivability. These include sending reporters to chase tornadoes across the
nation's heartland, follow the repair of the Hubble Space Telescope from
inside mission control, and dig for dinosaurs in Mongolia's Gobi Desert
(aka "The Middle of Nowhere"), using a satellite telephone to file stories
and pictures each day for a month.*

*What happens next is anybody's guess. I hope my next job is exciting,
fun, enlightening (for me and others), and contributes to society in a positive way. But for all I know, it doesn't even exist yet.*

BROADCASTER
(IN SOME CONTEXTS KNOWN AS "TALENT")

Science journalists who report their stories on the air or host a talk show,
like Joe Palca or Ira Flatow, are probably the most visible (or audible) members of the science broadcast community. There are few media outlets that
hire full-time science correspondents—NPR has what may be the largest
science broadcast staff in America. Reporters for local news shows may
find themselves reporting on science, and more often on health stories, but
that is rarely all they cover.

Ira Flatow, host and executive producer of *Science Friday*, National
Public Radio, says:

*I've been reporting on science for National Public Radio for more than 25
years. As a science correspondent from 1971 to 1986, I covered science,
health, and technology for NPR, reporting from such far-flung places as
the Kennedy Space Center, Three Mile Island, Antarctica, and the South
Pole. Since 1991, I've been the host and executive producer of* Science Friday, *a science talk show on NPR.*

*Before joining NPR, I was the news director at WBFO-FM in Buf-
falo, New York. It was at WBFO that I got my start as a reporter while
studying for my engineering degree at State University of New York at
Buffalo.*

In the midst of my career as a radio reporter, I also worked in television. From 1982 to 1988 I was the host and writer of the Emmy Award-winning PBS-TV science program, Newton's Apple. *I've also reported on science for* CBS News, CNBC, Westinghouse (PM Magazine), PBS, *and on health and medicine for* Newsweek Video.

EXECUTIVE PRODUCER

The executive producer is the manager of a particular media program. As the executive producer for *Newton's Apple,* Richard Hudson is the chief entrepreneur of its new science initiatives. His executive producer role requires him to cultivate relationships with the National Science Foundation and other sources of funding for the show.

In any given week, he must follow not only advances in science and the possible stories they might lead to, but also developments in science education, where television plays an increasingly important role. He spends a lot of time and energy networking with the science-education community, and because of that relationship he has been very involved in the development of the new national science education standards.

As the executive producer of *Science Friday*, Ira Flatow developed the idea for the show, lined up funding, and hired the production staff. He also negotiated contracts for rights, equipment, office space, telephones, and everything that relates to putting on the show. He continues to manage the staff and budget.

Richard Hudson, Executive Producer, *Newton's Apple*, KTCA-TV, Minneapolis, took a circuitous route to his current job:

I majored in physics in college, and worked for a commercial research lab for a couple of years after I graduated, starting grad work in solid-state physics at night. But the draft was still in force, and my number came up. I was an active musician, so I enlisted in the U.S. Army Chorus in Washington for 3 years. While there, I got a master's in drama (another interest) and, after leaving the Army, another master's and a doctorate in Music. My specialty was opera, and for a decade I stage-directed operas for regional opera companies, gradually working more and more on televised productions of theater and opera.

In Minneapolis, I started the touring company of the Minnesota Opera, and I had a number of contacts at KTCA-TV. They had just started a funky science show, with that well-known NPR science guy, Ira Flatow,

*and were looking for producers with varied experience and I fit the bill.
Around the* Newton's Apple *production schedule, I did a couple more operas and produced two holiday music programs for PBS, but gradually
devoted myself more and more to communicating science.*

*I now serve as the science editor of the many projects we have here,
and I am the executive producer of* Newton's Apple.

WEB DEVELOPER/PRODUCER

ScienCentral is currently producing the *Science Friday* Web page
(www.sciencefriday.com). For a broadcast journalist, there are special
problems and opportunities that come with producing a web site. Good
writing for the web needs to be short but clear, and supplemented by links
to other relevant web sites. Graphics and links make a web site stand out.
The medium is so new that there are few boundaries, and the sky's really
the limit. Interactivity with the audience make this kind of journalism very
immediate and direct.

Managing a web site requires the usual management and funding skills.
Since it is a brand-new medium, nobody knows how to make money at it
yet—so being a creative businessperson helps, too.

Eliene Augenbraun, web developer and executive producer at *Scien-
Central*, tells about her path from the bench to the Web:

*My career is typically atypical. During college, I worked as a lab techni-
cian while majoring in medieval literature, biochemistry, and behavior. I
went to medical school and became a medical illustrator. I worked as a
freelance illustrator and writer, and enjoyed it immensely, but realized
that I wanted more intellectual challenge and independence. Continuing
as a freelancer, I went to Columbia University and got a Ph.D. in biology.*

*I went on to do the usual academic thing—a postdoctoral research fel-
lowship. But I had some unique experiences that changed the course of
my career. In 1993, I was elected co-president of the Johns Hopkins School
of Medicine Postdoctoral Association, an organization I helped found. I
wrote our articles of incorporation and by-laws, got money for our projects, and reported for and edited our newsletter. I didn't know it then, but
it was the beginning of my science journalism business career. After my
research fellowship, I worked in a science museum for a year, then started
an American Association for the Advancement of Science (AAAS) science
policy fellowship in Washington.*

During my interview, I mentioned what I had done at Hopkins. To make a long story short, I ended up organizing a seminar on young scientist career issues for the annual AAAS meeting the following year. The seminar was a great success. I was invited to be a guest on NPR's Science Friday, *broadcast live from the meeting.*

After the show, I talked further with Ira Flatow, the show's host and executive producer. As a result of that conversation, and many more over the next several months, we started a media production company, ScienCentral, Inc. For the first year of ScienCentral's existence, I kept my day job at the U.S. Agency for International Development, where I evaluated media in new democracies. I am now working full-time as CEO of our company, developing business, managing budgets and people, doing web, TV, and other production work.

This company is the greatest experience I can imagine. I got it through the strangest series of lucky events, not through any grand plan. Yet everything I have ever done seems to have led me to this company.

Look for us, on a web page or TV near you . . . .

SO HOW DO YOU FIND THESE JOBS?

After reading this far you think you'd make a good science producer, reporter, or editor. But you're still not sure exactly how to make the leap from the lab to TV, radio, or the web. Here are some training resources that might get you on your way—they've worked for some of the people profiled here.

AAAS Mass Media Science and Engineering Fellows Program

This program, run by the American Association for the Advancement of Science, places scientists-in-training at media sites around the country for a 10-week summer program, giving them a chance to experience firsthand the difficulties and rewards of reporting science to the public.

Every year, from 15 to 20 scientists are assigned to work as reporters, researchers, and production assistants at radio and TV stations, newspapers, and magazines. News organizations that have hosted fellows include *The Dallas Morning News, Newsweek,* WOSU Radio (Columbus, OH), *The Oregonian*, KUNC Radio (Greeley, CO), *U.S. News & World Report, Time,* the *Milwaukee Journal Sentinal, National Geographic Television*, and the *Raleigh News and Observer.*

For more information about this AAAS program, including an application, write:

Mass Media Science and Engineering Fellows Program
AAAS
1200 New York Avenue, NW
Washington, DC 20005

Science Journalism Programs

There are several universities across the country that offer science journalism programs, and many more that offer science writing courses. Usually, broadcast journalism is part of the training, although the focus tends to be on print journalism. Some of the better known programs include those at Boston University, Columbia University, The Johns Hopkins University, New York University, the University of California at Santa Cruz, the University of Missouri, and the University of Wisconsin-Madison.

A 1996 directory of science communication programs and courses is available from the University of Wisconsin-Madison. You can order the directory by sending a check for $10, payable to the School of Journalism, and sending it to:

Sharon Dunwoody
School of Journalism and Mass Communication
University of Wisconsin-Madison
821 University Avenue
Madison, WI 53706

Professional Societies

Professional societies are another good source of information on science journalism as well as broadcast journalism. The National Association of Science Writers (NASW) publishes a quarterly newsletter, and has a web site that includes job listings (www.nasw.org). For more information, write, call, or e-mail:

NASW
P.O. Box 294
Greenlawn, NY 11740
71223.3441@compuserve.com
(516) 757-5664

There are local chapters—ask the national organization for information about one near you.

Internships

Finally, if there's a radio or TV show, or a media organization or production company that you're interested in, consider an internship. Some places have more formal internship programs, while others may never have had an intern. Whatever the case, make some calls and find out the name of the executive producer, producer, or editor, and then offer your services (free of charge, usually). If you can muster up the grace to handle even the lowliest of tasks with a smile (all the while thinking, "Did I go to college, grad school even, to make copies?"), you may find your hard work rewarded with more challenging assignments.

Whatever you end up doing, keep in mind that you'll be getting a rare inside look at what goes into making a TV or radio show, or a great web site, without committing yourself to more than a few months' work. That should be enough time to figure out if there's a job there you'd like to do, and enough time to make some contacts that may come in handy when you're looking for a real job.

In conclusion, we leave you with some advice from NPR science correspondent Joe Palca. He says, "Don't be constricted by someone else's idea of what a successful career should be. Science gave me the skills to climb a different ladder. Find your niche. And if you have the journalism bug, then go for it. There may not be that many jobs in journalism, but there may be more than in science."

6

· · · · · · · · · ·

PITUITARIES TO PINSTRIPES:

A Path to Venture Capital

· · · · · · · · · · · · · · · · ·

Andrea Weisman Tobias, Ph.D.

Associate Director, Abingworth Management Ltd.

I am writing this chapter in my office on St. James's street in London where I am an associate director at Abingworth Management Ltd. (AML). AML is a British venture capital firm that invests in start-ups in the international biotechnology industry. St. James's is a far cry from my graduate student days at UCSF, when I made weekly trips to the local slaughterhouse in Berkeley to gather fresh pituitaries from cow skulls for my doctoral studies. Not a memorable experience, even at the best of times.

So how do I find myself in London, in a Maxmara business suit having tea at the Ritz with my colleagues to discuss how to invest our new $100 million fund? It has been an interesting journey from lab bench to Green Park, so I will begin.

First, I would like to state that almost anything can be achieved in this quickly changing, entrepreneurial world if one has good credentials, a clear vision, flexibility, and optimism. Science itself has no cultural or linguistic barriers, and a Ph.D. is truly an international credential. Physicians and lawyers are constrained geographically by their licenses to practice their profession but scientists have the freedom to live and work around the world. Also, being fluent in a foreign language makes a résumé more interesting and is a definite plus in our global work environment.

How I Began My Trek

My first dream as a full-fledged scientist was to live and work in Paris, France. I had been inspired to work in Paris by a well-known neuroendocrinologist when I was an undergraduate. Imagine my extreme disappointment when I learned that after finally completing my Ph.D. 5 long, hard years later, I still lacked the experience that she required to work in her Parisian laboratory.

But now I was even more determined to get to Paris one way or another. I accepted a 3-year postdoctoral fellowship at McGill University in Montreal (a French-speaking city!) with the promise of spending my final year in Paris. My senior professor was good to his word, and the year that I spent in Paris in a molecular biology lab at Roussel Uclaf (now Hoescht Marion Roussel) was a turning point in my professional life.

Although I returned to an academic atmosphere for another 2 years at the Collège de France in Paris (my original inspiratrice finally invited me into her lab as a visiting scientist), a future in industry now intrigued me. But I wasn't quite ready to make the leap and I seriously considered remaining at the Collège, or accepting an academic-track position at Georgetown University in Washington D.C. Life being unpredictable, personal events led me back to the San Francisco Bay Area where I was bitten by the biotech bug.

After performing research for 12 years, I knew that at this point in my life I wanted to get off the bench. But now I was in a real "Catch 22"; no one wanted to hire someone for an "off the bench" position unless you had experience off the bench!

With no other obvious alternatives, I was compelled to take a job as a cancer researcher at Triton Biosciences, a small biotech company founded by Shell Oil Company and later acquired by Schering AG to become Berlex Pharmaceuticals. Athough I was still tethered to a lab bench, I made every possible effort to learn about other aspects of the biotechnology industry.

I did manage to help organize external collaborations related to my research, and I acquired some knowledge about intellectual property and clinical trials. At this time, I also learned the importance of getting to know people outside of basic research to broaden my horizons. I must add that sometimes both timing and luck play a role in defining a career path and I have been the beneficiary of both.

Still on the bench a year and a half later, I picked up the phone and called Genentech out of sheer frustration. I inquired if they had any "off the bench" positions that required someone with my skills and international background. Destiny was clearly on my side, because Genentech was just in the process of creating a collaborations program. After 11 grueling hours of interviews, I was the first manager hired.

My job as "manager of new research identification" involved attending scientific conferences and arranging academic collaborations around the world, organizing scientific roundtables, and writing comprehensive reports on new research areas of interest.

This was my dream job. I was given the opportunity to attend Genentech's research review committee, whose members included the VP of Research, the VP of Clinical Affairs, key senior directors, and world-renowned external consultants. These discussions were enormously educational, because all research-based decisions were made at these meetings.

I also routinely met with department directors and scientists to discuss new ideas, as well as potential and on-going collaborations. Although I had little interaction with business development, I spent time in the legal department working on collaborative agreements, patents, and intellectual property issues. I also worked with marketing while preparing my in-depth reports on various business opportunities presented as scientific programs. As the Genentech liaison to both the Canadian and Japanese offices, I was obliged to develop good communication and people skills.

Later on, when I devoted the majority of my time to the cardiovascular department, I had to earn the respect of the scientists before I was finally accepted as a true member of the department. The greatest hurdle in my unique position was convincing my colleagues that my brain did not atrophy despite the fact that I no longer held a pipette. Eventually, I succeeded in becoming a respectable "armchair scientist." My job was very rewarding and only a major reconstruction of my department compelled me to move on.

Once again, I was at another crossroads in my life and career. But this time I was better equipped. During my years at Genentech, I had the opportunity to develop strong people skills due to the sheer number of diverse people that I interacted with on a daily basis. I also discovered that I had a "nose" for new opportunities; I had been responsible for establishing a variety of academic collaborations in the areas of allergy, angiogenesis, cancer, and even snake venom (antithrombotics).

I was offered the opportunity of remaining at Genentech and writing in-depth reports full time or taking a lateral position in the marketing department. But I did not believe that either of these positions had enough growth potential or fully utilized my skills.

During my job search, I was eventually offered six external job opportunities and all were different. These offers were primarily the result of personal contacts and the fact that my experience at Genentech, a well-known and world-class organization, was invaluable. The positions were in intellectual property, venture capital, strategic development, or business development in small biotech companies, large biotech companies, a well-known investment banking house, and a technology transfer office at Oxford University. I knew that I was in an enviable position because I had so many choices.

After a few sleepless nights, I chose to become a director of strategic development in Chiron Corporation because I felt that this was the best fit for me both personally and professionally. I have never regretted my decision, as it was an excellent learning experience that broadened my business development, operational, and analytical skills, as well as bringing me a personal bonus—I met my husband in the office next door!

My position as director of strategic development included a diverse range of responsibilities and tasks. I created and chaired a technology assessment committee (a team comprised of scientific directors and business development representatives from all divisions) that evaluated and built databases of all unsolicited technologies and collaborative opportunities (600–700 opportunities each year). I led large and small task forces and strategy teams, and contributed to strategic plans in specific research areas. I organized the first scientific retreat as well as innumerable scientific and corporate meetings, performed in-depth due diligence with patent attorneys for select projects, and even dabbled in negotiating licenses and writing termsheets. I was also the scientific liaison to Ciba-Geigy after it had acquired almost 50% of Chiron.

I was extremely busy! During the 3 years that I spent at Chiron, I had the opportunity to meet with approximately 100 biotech and pharmaceutical companies, which gave me an excellent base of professional contacts.

During this time, I had been watching the Biotech industry in Europe finally gain momentum, especially in the United Kingdom. In the back of my mind, I always had the goal of returning to live and work in Europe. So with this in mind, my husband and I chose London as our first choice because of the language, the city, and the exciting job possibilities in the health care industry.

Now, here is where tenacity, good luck, and personal contacts played a role once again in my life. I was given the names of key contacts (David Leathers, a senior partner in Abingworth, George Poste, head of R&D at SmithKline Beecham, and Tony Scullion, the worldwide director of business development at Glaxo Wellcome) in venture capital firms and pharmaceutical companies in London. Since I had already lived in both Canada and France and had managed to obtain work permits in these countries, I was not easily daunted by the naysayers (and there were many) that said it could not be done.

Upon receiving my résumé, Dr. Stephen Bunting, the other senior partner at Abingworth, actually telephoned me the next day. Abingworth was seeking a technical consultant and they liked my American biotech and international background. I eventually left Chiron, worked for Abingworth as a freelance consultant while living in the Bay Area, and relocated to London in early 1997. One major word of advice is to NEVER listen to anyone who tells you what you cannot do—everything is possible with solid credentials and the right attitude.

WHAT IS A VENTURE CAPITALIST?

So what does a venture capitalist actually do?

In basic terms, a venture capitalist must first raise capital to create a fund, then invest it, create liquidity (by selling shares in portfolio companies), and return cash to the fund shareholders. This process involves careful evaluation of the opportunity itself (in this case, biotech companies), management of the investment once it has become a reality, often taking a seat on the board, and ultimately making the company a success. In the VC world, real success means a 5- to 10-fold return on the original investment. A real VC may also sit on three to four boards at any given time.

To give a few specifics, Abingworth was founded in 1973 and is one of the largest independent venture capital groups in Europe, as well as one of the earliest biotech investors in the United States. AML currently has $250 million under management (85% is invested in biotech, including the new $100-million fund). AML has invested almost $100 million in 34 biotech companies (4 in the U.K.) over the last 10 years. Twenty-four have gone public and 5 have been acquired. The new fund will be invested in 15 to 20 new companies and a few public companies, with some emphasis on investing in the United Kingdom and Europe. AML has three directors (one chartered accountant, one Ph.D., and one M.D.), one associate director (myself), and one in-house consultant with many years of experience in the biotech industry.

A typical day in the life of a VC (which in London begins later and ends later to stay in sync with the U.S. stock exchanges based in New York) involves listening to presentations by new or established companies, speaking with consultants, checking stock prices, keeping up with industry news (this requires extensive reading), and making innumerable phone calls to CEOs, potential management hires, other VCs, investors, lawyers, consultants, analysts, and so on.

Phone calls can actually take up 20% of every day! In addition to all of the above, our firm has weekly biotech meetings in which all potential and ongoing investments are evaluated and discussed. Members of the team submit written proposals and must defend investment recommendations to other partners. Although unanimous agreement is preferred, at least two out of three directors must be in favor of moving forward before an investment is made.

What Makes a Good VC?

Interpersonal, creative, and analytical skills all contribute to one's success as a VC. A VC may have to convince other VC organizations to co-invest in a new opportunity and come up with a creative financing that pleases all parties, a VC has to be sensitive to the needs and biases of the academic scientists who are company founders, or sometimes VCs may need to influence and persuade board members to pursue a strategy that is best for the development of a nascent company.

Since business and negotiating skills are a must, a scientist can obtain some of these skills either by working in a business development, licensing, or legal department in industry, or in a technology transfer office in academia. Accounting skills have a limited role in the venture business, and a basic finance course is probably adequate background.

Being a VC is generally a very lucrative position and salaries are in the six-figure range. Overall compensation is much higher (it can be above 7 figures) because VCs receive 20% of the capital gains of the fund over the life of the fund, which is usually 10 years.

But it can take from 7 to 8 years before a VC sees major financial returns. The training phase averages 5 years, but it can be as little as 3 years. It requires the following attributes or milestones: strong people skills, decision-making skills, the ability to identify new ideas and to create opportunities around them, the capacity to attract financing from investors, good problem-solving skills, and the ability to lead.

An extrovert with strong people skills probably has the easiest chance of succeeding in this profession. However, the job is so broad that if someone has some but not all of the attributes described above, he or she can still succeed. The promotion ladder from lowest to highest usually consists of adminstrators, a manager level (lesser academic degrees and work experience), a director level (nonpartner, often technical, more responsibility than managers) and full partner level, although each firm differs. Partners (termed directors in the U.K.) decide who and when someone will be promoted to join their ranks, and this process differs among firms. Junior partners may have a wide variety of backgrounds including business, finance, medical, or scientific degrees. Interestingly, industry experience is not always a prerequisite.

As an associate director, my job involves conducting in-depth due diligence on early-stage companies or "seed" deals, where the actual company may not yet be formed. This includes analysis and assessment of the specific scientific platform, sector (competition), intellectual property, and management team. I frequently submit a written evaluation or proposal to the team for discussion at weekly biotech meetings and sometimes lobby partners prior to a decision-making meeting. Funding recommendations usually range from $2 to $5 million or greater over the life of the investment.

My role also involves assisting companies in our portfolio in a number of ways. For example, I may give advice regarding specifics of the scientific program, contribute to strategic planning, help to find corporate partners, or even help to recruit additional talent. Finally, I try to identify innovative new research areas and/or founding scientists from which new companies can be created. Good communication with colleagues and experts in the field, extensive reading, and attending scientific and industry conferences are important activities in this position.

There are a limited number of VC organizations (especially in the U.K. and in Europe) and the profession has a "clublike" atmosphere. The firms are usually flat organizations and they have a finite number of job slots. Available positions are rarely, if ever, advertised, and job openings are communicated by word of mouth or by professional recruiters. An ex-VC may become a CEO or COO of a start-up or an established company, or he or she may even go into politics!

Among the enjoyable things about being a VC are that one can meet interesting, accomplished people and can learn something new everyday. There is a great deal of travel and freedom, each workday is varied and stimulating, and the job is financially well-compensated.

Attending to administrative details (legal documents) is probably the most tedious component of the job. Being a VC requires extensive travel, often as much as 40 to 50% of the time. Also, being a VC usually means wearing formal business attire, literally and metaphorically, and this is something that scientists may not particularly enjoy. Decision making, which is a 6- to 8-week process on average (although there are certainly exceptions) in the VC world, does not require the depth of understanding or proof of principle that scientists are constantly striving for, and this can take some adjustment. The stress level is fairly high, but rarely unbearable.

There are many different paths that can lead to becoming a VC. A Ph.D. gives a scientist the credibility that is often required for a first real job or big break in any industry. Formal scientific training has been absolutely essential in my own career path, which often involved in-depth technology assessment and discussion with world-class scientists in a broad range of research areas. My technical background was vital in enabling me to enter the venture capital profession. Also, timing and a few key personal contacts are invaluable in pursuing and developing any career path.

In conclusion, a good scientific foundation can open doors to any number of rewarding jobs in research, business, finance, or informatics, in the rapidly expanding health care industry. Personally, I am a strong believer that if one has good credentials, vision, courage, and tenacity, almost any career goal can be achieved.

How I Became an Analyst:

Science-Based Investment Advisor

· · · · · · · · · · · · · · · · ·

Mary Ann Gray, Ph.D.
Biotechnology Analyst, SBC Warburg Dillon Read, Inc.

THE EARLY YEARS

Even as a child I was interested in how things work. When I was six, my father got an electric razor for Christmas. That afternoon, my parents found me in my room with pieces of the razor scattered around. When asked what on Earth I thought I was doing, I replied, "I just wanted to see how it worked." Over time, my interests evolved into an interest in how the human body works. Living in South Carolina from age 11 through college, I was never really exposed to scientists. Consequently, I thought that the only way to delve into the mysteries of the body was to become a physician. Since I was very shy, the thought of dealing with lots of strangers was not appealing and I became attracted to pathology (dead people are a lot less threatening!) and I settled on becoming a pathologist as early as seventh grade.

High school was not a wonderful experience for me and I decided to leave high school after my junior year and enter the University of South Carolina as a non-high school graduate in the summer of 1969. Of course, my major was biology.

After my freshman year, I got married and had a child at age 18. Needless to say, this caused some changes in my plan to attend medical school and become a research pathologist. After my son was born, I returned to school right away, but had to attend at night so that my mother could babysit while my husband worked evenings. Back then, science courses were not offered at night so I worked on my electives for about a year and a half. When my son was two, I went back to school full-time and I graduated in December 1973.

BEGINNING A CAREER

Medical school seemed out of the question at that point, so I did the next best thing—I become a medical technologist at Moncreif Army Hospital in Fort Jackson, South Carolina. This allowed me to become familiar with the general workings of a laboratory as I rotated through the various departments. I also got to deal with patients when it was my turn to go on the floors and draw the early morning blood samples. More importantly, I got to work with pathologists. When there was an autopsy to do, most of the technologists became very busy with other things. I, however, was always eager to help. Our pathologist was a wonderful teacher, and I learned a great deal about anatomy and pathology during these sessions in the morgue.

Soon I became bored with the routine of the hospital laboratory and decided to try to find a job doing "real" research. I moved to Northern Virginia to be near the National Cancer Institute and I started working for a contract research firm. Our specific task was to manage the preclinical toxicology of new anticancer drugs being developed by the National Cancer Institute. We did not perform the tests ourselves, but selected various subcontracting labs and then we provided an independent assessment of the results, including hematology clinical chemistry and pathology reports.

This job was one of the many turning points in my career. I was given as much responsibility as I could handle and I was supported by my boss, an excellent toxicologist. I also worked hand-in-hand with scientists at the NCI and at various pharmaceutical companies and universities. It was with the help of these wonderful friends and mentors that I gained the courage to go back to school 7 years after receiving my bachelor's degree.

Since I was a little uneasy about returning to school, I chose to attend the University of Vermont. At that time, Vermont had the characteristics that I was looking for in a graduate program, and two of my former colleagues were on the faculty. The pharmacology department was small, but Vermont had an excellent regional cancer center, and the laboratory and clinical staff interacted a great deal. I felt that I would be able to learn how to gear my research toward applications that would be important in patient treatment. I was not interested in true basic research, but in transplanting the principles discovered by the basic scientists into patient care.

My graduate school project evolved from my previous work with the anticancer drug doxorubicin and its analogs in an attempt to find a drug with reduced cardiotoxicity. My focus was the potential mechanism of this cardiotoxicity and ways to reduce it, and I developed models to screen new compounds for improved characteristics. I continued to pursue this research during a year of postdoctoral research at the Northwestern University Medical School and then I expanded this work to other anticancer agents with Dr. Alan Sartorelli at the Yale University School of Medicine. I had never aspired to an academic career, and after 2 more years as a postdoc, I decided that it was time to make my break into the corporate world.

THE BUSINESS OF SCIENCE

Because of my experience with antitumor drugs and my expertise in developing animal models, I was hired by a private biotechnology company to preside over the preclinical development of certain monoclonal antibody conjugates.

The biotechnology world was fascinating in a number of respects. The work was focused and the goals were clear. The company had approximately 70 employees when I joined and the atmosphere was very collegial. We were all pulling together toward the same end, getting a product on the market. Project management meetings exposed us to input from all areas of the company. I worked with clinicians to design animal models that would be better predicators of results in humans, I learned what things were important to our fledgling marketing and sales group, and I learned how to manage both a budget and people. Wearing many hats in a small organization is a wonderful way to gain at least some appreciation of all aspects of building a business.

During my time at the biotechnology company, management decided to pursue an initial public offering (IPO). This was a fascinating experience for me, since my involvement with Wall Street until that time was limited to a brief scan of the business pages in the local newspaper. It became even more interesting when we tried to close the deal in October 1987, the day of the big stock market crash. Not surprisingly, the deal was postponed until a better time.

After a few years, the fortunes of the biotech industry and this company in particular diminished. The products took longer to develop and were more expensive than expected. We went through one downsizing that was extremely painful, especially in a small company where all of the employees knew each other fairly well. Even worse, the first cut did not achieve the goals and there was a second layoff only a few months later. The moral of the remaining employees was at an all-time low. I decided that it was time for me to move on.

In considering my options, I was looking for something new. I had done the academic thing, contract research, hospital laboratory, biotech company. What was missing? Aha! A major pharmaceutical company was something I had not yet tried. That decision led to a move to Schering-Plough in the relatively new tumor biology department, where I was in charge of the preclinical work on potential new treatments for cancer.

Most people make a move from the pharmaceutical industry to the biotechnology industry, not the other way around—and now I know why. At a smaller company, everyone tends to work together. People wear many hats and you are expected to voice your opinions to help keep projects on track. It is a simple matter to talk to the vice president of research or even the CEO whenever you wish to express an opinion.

In a large pharmaceutical corporation, the organization and team approach does not usually kick in until lead compounds are selected and handed over to the clinical development group. Research is much less directed, and I felt as if I were spinning my wheels. I felt that I was moving farther away than ever from my goal of developing products to treat patients.

THE RIGHT PLACE AT THE RIGHT TIME

As I was trying to decide what my next move would be, I talked to a friend who had recently attended a meeting where several Wall Street biotechnology analysts had given presentations. She told me that it sounded like the ideal job for me—they get to work on a variety of projects (hence no boredom), they travel (I like travel), and they get to interact with a diverse and interesting group of people (I have definitely outgrown my shyness).

It sounded interesting, so I begin to write letters to the analysts asking for information. Most important for me was whether they thought it necessary that I obtain an MBA before I could move into an analyst slot. Having spent a good portion of life in school and having just finished paying for my son's college, I was not anxious to tackle another 2 years of school and, more importantly, to lose income.

I received several polite replies, but in December 1991 I received a call from the analyst at Kidder Peabody. This analyst had a Ph.D., but also had an MBA. At this time, the biotechnology market was really "hot" and investment banks were working on numerous biotech IPOs. Evaluating all of the existing and potential companies was too much for one person, and Kidder Peabody needed help. Again, I asked about the need for an MBA and was assured that the scientific expertise was more important at this point. (I have also come to realize that my experience working in the both the biotechnology and pharmaceutical industries is equally important.) In February 1992 I left Schering-Plough for Wall Street.

During my first few months, I was meeting with companies and evaluating technology nonstop. However, things quickly came to standstill when numerous clinical disasters caused the industry to fall out of favor with investors. On the positive side, this lull gave me an opportunity to get my feet wet without too much intense pressure. It was a good thing I got this time to come up the learning curve, since 6 months later the senior analyst left and I was on my own.

After about 3 years, a change was forced upon me. In late 1994, PaineWebber, another investment bank, bought Kidder Peabody and brought in their own biotech analyst —suddenly, I was out of a job. While in the past I have always been ready to move on after about 3 years, this time I wasn't even close to getting bored. Had I finally found the right job? And how could I stay in the game despite the merger?

Because being an analyst is a little like being an independent operator, I decided to see if I could use my experience and contacts to run a consulting service. I formed BioLogic, Inc., and began to work on a broad range of projects for numerous clients. These projects included investor relations strategies, writing private placement documents, providing valuations for potential acquisitions, and writing a newsletter.

After a successful first year, I was getting to the point where I needed to hire a full-time assistant. I had never before been responsible for paying someone else's salary, and it made me a little nervous. As I was wrestling with this decision, I received a call from a former Kidder colleague, now at Dillon Read. (The majority of Kidder's health care team ended up at Dillon Read.) Would I like to come back to Wall Street? I must say, the answer would probably have been no if it had been anyone else asking. But with this group, I knew what I was getting into and I knew that it would work as it had in the past. I have now been at Dillon Read a little over a year and I have just learned that we are being bought by SBC Warburg. (See, I don't have to keep changing jobs now, my job keeps changing on its own!)

WHAT IS A BIOTECHNOLOGY ANALYST?

In general, the job of a biotechnology analyst is to evaluate biotechnology for investors. The bottom line is whether an investor can make money by investing in a particular company. The trick is to figure out which of the many companies represent the best investing opportunity—and it's not always just a question of which company has the best science.

Many people think that we have armies of analysts for each industry, but this is not true. Most firms now have two analysts that may be supported by one or two research associates. With more than 350 publicly traded biotechnology companies for investors to choose from, the task of selecting the best investment opportunities with limited resources can be daunting.

I think of my job as having three major parts: research, marketing, and corporate finance. At various points, the time spent on each part can vary a great deal. One common thread is that there is never enough time. In this fast-paced business, you always feel that something is being neglected.

Research involves a number of activities, including company-specific research and industry research. When evaluating a company, you must look at the products they are developing, the technology that they use, the management team, and their business strategy. The first source of information is usually the company, but facts need to be verified by independent sources.

This is the area that is most closely related to scientific research, where scientific skills are readily useful. The type of information that you are evaluating may be in the form of scientific publications, abstracts, or company presentations. Keep in mind that the goal here is to eventually make money for investors, so no matter how scientifically elegant a project may be, you've got trouble ahead if the ultimate product is not clinically useful, is too costly, or is otherwise unsuitable for use in a patient population that is large enough. It is also important to have a broad scientific background. The technology that you are evaluating will not always be in your area of expertise. I was trained as a classical pharmacologist and this broad background has been useful in evaluating information from areas outside the cancer field, such as neuroscience, virology, and the like.

There is never enough time to evaluate these areas and you can't possibly do enough reading to quickly become an expert in an entirely new field. Therefore, it is important to develop good instincts and a strong network of scientific and medical colleagues on whom you can depend.

I also try to attend several scientific and medical meetings each year. The medical meetings tend to be more useful, since I can get a feel for the state of the art in the treatment of various diseases and what technology physicians are excited about. These meetings also give me a chance to speak with experts in various fields about technology that I am evaluating. When data from one of the companies that I follow is presented, I am always very interested in the questions asked by the experts in the audience.

In addition to evaluating the scientific feasibility of a company's products, you must also evaluate the market potential. How many patients have the disease? How are they treated now? Will the new product be useful in all patients or only in a small subset? Is the new product better than existing therapy? Will doctors use it? How much will it cost? Will insurance companies and HMOs pay for it? These are questions that most scientists are not used to asking, but they are crucial to evaluating a new product. It is important to think practically and to understand how physicians treat patients.

Okay, let's assume that the company has a potentially great product that looks like it is going to work, physicians think it will be useful, and it saves money so it will be reimbursed. Does management have a plan to get the product into the marketplace and the experience to execute that plan?

This is one of the more subjective parts of the job, but it is also very important. The managers of a biotech company have to play a lot of roles and they must communicate with a diverse audience. I try to rely on track records as much as possible and I follow a company's progress for awhile before making a "buy/don't buy" decision. This requires numerous meetings with management to define their goals, following up on progress over time. Are they overly optimistic about timelines and costs? Is there always an excuse for not achieving milestones? Or do they always meet their stated goals on or ahead of schedule? I must also admit that gut feelings play a role, but it is crucial to develop a trust in management.

Okay, so assume that the company has products in the pipeline and they've put together a great team. Now we need to look at whether investors funding this enterprise actually make money. For most biotechnology companies (with only about 6 exceptions), profits will be made in the future. Trying to project future earnings, much less to determine their value today, is a difficult task that involves some fortune-telling skills. Public companies rely on investments from the public in order to bring their products to market. This is a costly enterprise that requires $250 to $450 million and usually requires from 10 to 12 years for just one product. What is the company's plan for financing this development process? How will management balance investment in research and development of new products with making a profit for investors as soon as is feasible?

After all of this analysis has taken place and you have decided that the company is a winner, you must explain how you arrived at that conclusion to a diverse audience composed of some very naive laypeople, some very sophisticated institutional investors, and everyone in between. You must write a report that conveys the important points without losing the attention of people who don't have a lot of time to read through your pithy prose. You must tell the story of what the company does clearly and you must also explain why you feel that it is the right time to invest. Investors need to know what likely events will increase the value of the company and the stock price over the next 6 to 12 months. (This is a long time to an investor.)

The companies that you "cover" include those for which you are actively writing reports and notes to investors, those for which you have a detailed financial model, and those for which you have made an investment recommendation. This recommendation is usually to buy, sell, or hold, or a variation thereof depending on your firm (different firms may use different terminology to avoid saying "sell"). A single analyst typically can cover from 10 to 15 companies on a regular basis, depending on the amount of support that he or she has.

All of this research leads into the second major aspect of the sell-side analyst job, which is to market these stocks to buyers. (Buy-side analysts work directly with portfolio managers and institutional investors to decide which stocks to buy.) Once you have these great investment ideas, you must communicate this enthusiasm to the investors so that they will buy (or sell) the stock.

The first line of communication is your firm's institutional salesforce. These people have specific client institutions, with whom they are in daily contact with buy, sell, and hold advice across all sectors. Your ideas are transferred to the salesforce via your written reports, morning notes (brief comments on the impact of specific news events on stocks), or other specific comments—a change in earnings estimates or something that has changed your opinion of the stock—on companies that you officially cover.

In addition to the written reports and notes, you must also talk to your salesforce. This takes place at a daily meeting at approximately 7:30 A.M. East Coast time (4:30 A.M. for West coast analysts calling in)—hence the term "the morning call." The meeting takes place so early so that the information that is relayed by the analysts becomes the subject of the salesforce's daily calls to their institutional accounts. This information must be relayed to the clients before the stock market opens at 9:30 A.M. (6:30 on the West coast) to allow clients to take action immediately. In addition to having the salesperson call the clients with news, it is also important for the analyst to call certain key clients directly. It is important for an institutional investor to get to know and trust the analyst who is providing the information on which the investor may risk a great deal of money.

Research on general trends in the industry is also important. What are the current areas of interest (or disinterest) for investors? What kinds of companies are paying off now? What characteristics do successful companies have in common? How do investors define the profile of the ideal company? (This changes frequently, depending on how well past investments are doing.) Is biotech a worthwhile investment and how do you chose the right companies? Written reports on industry trends and fundamentals are helpful for investors, but they also help you to formulate and articulate clear, concise opinions about your industry.

In addition to written reports, morning notes, and discussions with the salesforce and clients, it is important to visit certain clients on a regular basis. Face-to-face meetings help build the relationship and generate trust. Clients can also be a valuable source of information. A typical marketing meeting can be a valuable two-way exchange of information.

Corporate finance is the third portion of the job. The corporate finance group, also called bankers, helps client companies raise money, advises them on mergers and acquisitions, and provides crucial strategic advice. Depending on the firm, the analyst may or may not be heavily involved in certain corporate finance decisions.

Initial public offerings tend to require the most work. I like to meet with, and get to know, interesting private companies early. This allows me the time to make some judgments about management and to follow the company's progress even before they become banking clients. It also gives me and the bankers a chance to guide the company through the process of deciding on the right time to approach the public markets and to decide what their other alternatives might be.

Once a decision has been made to go forward with a public offering, I tend to participate heavily in the "due diligence" process. This involves numerous conference calls with collaborators, scientific founders, corporate partners, physicians, and others to evaluate the company. This process is somewhat akin to checking references. I also go through all of my typical research steps, including building a detailed financial model of the company.

After due diligence is completed, the company must write a prospectus, a document filed with the SEC that is given to potential investors. The business section of this document describes the company in detail and it is important that this section be clear and understandable. I like to take part at least in the initial stages of drafting this document. The next step is to prepare a 30-minute slide presentation that will be used to describe the company to potential investors around the world in a grueling ordeal called a road show.

During the road show the company will give this presentation numerous times in several key cities throughout the United States and Europe. Some of the presentations will be to groups of investors over breakfast or lunch. Other presentations will be one-on-one meetings with key investors. I like to have some input in the slide show, but I do not typically go on the road show. I do make follow-up calls to investors who meet with the company so that I can answer any questions and describe why I feel that the company is a good investment. Once the offering is completed, this company is added to my coverage list and I write a detailed research report.

Other activities related to corporate finance include follow-on offering for existing public companies and evaluating potential merger or acquisition candidates for our clients.

POSITIVES AND NEGATIVES

As you can tell, there is never enough time to do everything as thoroughly as you would like. This keeps the job exciting, but it is sometimes frustrating, especially if you are a scientist who is used to examining every last detail. The ability to disseminate information rapidly is important and has to be balanced with a well-thought-out opinion.

The job requires working long hours. The morning call is at 7:30 A.M., and if you haven't written a morning note on a topic the night before, you have to come in early enough to have it written and approved prior to the call. The evenings are often a time to catch up on reading after the phones have stopped ringing, and frequently there is a dinner meeting with the managers of a company that lasts until 9 or 10 P.M. In addition, visiting companies, attending scientific and medical meetings, visiting clients, and attending and presenting at industry conferences require a great deal of travel.

The job requires a great deal of dedication because of the hours and the amount of energy spent to keep on top of rapid advances and competition. On the positive side, it is never boring. If you grow tired of working on a medical topic, you can focus on a financial model, or you can call a client to chat about the industry.

Job security is nonexistant. Everytime the biotech sector tanks, many banks shed their biotech analysts. They rehire when the sector heats up again, but this is not much consolation. Also, as I experienced, the merger mania in the banking world can throw a monkey wrench into your career plans.

How much are analysts paid? This is a tough question to answer because pay can vary greatly and rumors abound. In general, analysts receive a base salary plus a bonus. The base salaries range from about $30,000 to $60,000 for a research associate, $60,000 to $90,000 for a junior analyst, and $90,000 to $125,000 and more for a senior analyst. The base salary is similar to that for most other types of jobs, and there is a range for each level. The names of the levels vary among firms, but typically the grades include research associate, associate analyst, vice president, senior vice president, and managing director.

The bonus is the more difficult part to figure out. It depends on many things and most firms do not have a preset formula for determining the bonus level. Also, different portions of the job can be more or less important at different firms. The bonus pool—the amount of money the firm dedicates to bonuses—is influenced by how profitable the firm has been that year, so bonuses can vary greatly from year to year.

The size of an individual analyst's bonus depends on things you can't control, such as how well the investment banking business in the sector is doing that year. Investment banks make a lot of money by taking companies public and conducting follow-on offerings, so analysts who works on several offerings usually gets a small piece of each fee as a part of their bonus. Some firms emphasize the rankings published in *Institutional Investor,* a publication that annually picks the top three analysts in each sector according to votes from institutional investors. Some see this as a popularity contest, but it is designed to pick those analysts who help institutional investors make wise investment choices.

Other factors that determine the bonus include quantity and quality of reports published, the opinion of the salesforce as to how helpful the analyst is to them, and the performance of the analyst's recommendations regarding their covered companies.

Bonuses vary from about $50,000 to $1 million for senior analysts, but in general they are at least several hundred thousand dollars. This makes an average analyst's compensation in the range of $300,000 to $500,000, but there are always many exceptions.

However, there are some who believe that good analysts are paid whatever amount it takes to get them to stay with the firm. This creates an atmosphere much like professional sports, where analysts often jump ship to a higher bidder. Rumors of $1 million signing bonuses and multi-million dollar contracts abound.

Writing is important. You must be able to write in a style that is easily understood and that conveys the gist of very complex issues in a way that holds the readers' interest. You also must write quickly, for time is, in fact, money in this business.

Biotechnology is a diverse industry, both in the technologies involved and the people. We can all appreciate the differences between a scientist who is happy in the lab with no one to bother them and a CEO who has the gift of gab. As an analyst, it is important to be able to walk in many different worlds and to communicate in different languages. It is the combination of these different worlds in the biotechnology industry that I find fascinating.

As scientists, we are taught to question, and that is also a key attribute for an analyst. Even when you listen to presentations outside of your field of expertise, your scientific training should allow you to evaluate the information that is presented and ask the right questions. In addition to my scientific training, my work in both the biotechnology and pharmaceutical industries has provided me with the experience necessary to ask the right questions regarding the business aspects of the company.

How to Become an Analyst

Unfortunately, the number of positions for analysts is limited. If you consider that most firms have on average two biotechnology analysts, there are probably no more than 100 biotechnology analyst positions. There are more junior analyst and research associate positions available, if you are willing to work for someone else so that you can learn the ropes.

Analyst jobs can be difficult to find. They are not normally advertised in newspapers or in other sources. In addition, research directors usually look for a rare individual with a variety of skills and perhaps some experience in the financial world. In order to demonstrate that you have the necessary skills, you should be willing to be flexible in your expectations.

One way to prove yourself is to act as a consultant to an analyst on a special project. You will be paid, you will gain experience in research and writing with the investor in mind, and you will have a product, that is, a report, that you can use as an example of your work.

You may have to accept a position as a research associate in order to learn the business. These jobs are not glamorous and they involve long hours of digging out information for the senior analyst. You could also write some research reports on your own. Pick a company or two that you find interesting, describe the investment potential, and build a financial model for the next few years. This provides you with material to present in an interview that demonstrates you writing skills and your analytical ability. In addition, you can call the investor relations department at the companies you chose to write about and ask them to send you some of the analyst's reports. This will give you an idea of what analyst reports are like.

Because the jobs are not advertised, the best place to start is with current biotechnology analysts. Go to the library and ask for *Nelson's Guide to Institutional Research*. This book lists all of the investments firms and it gives the names and addresses of all of the analysts. Begin a letter-writing campaign. You can start by requesting information and you can follow up with phones calls for more information: remember that analysts are busy and they may not return your calls. Be persistent. The key is to be at the right place at the right time, so send your résumé out periodically (every 6 months or so) even if you have not received a response or if you have received a negative response. The landscape is constantly changing as analysts move around and needs change.

There is not really a corporate ladder to climb, unless you start as a research associate or a junior analyst. A research associate helps the analyst do the digging, writes drafts of reports and morning meeting notes, and works on financial models. Research associates also perform certain administrative functions at some firms. Junior analysts typically have a number of small companies that they cover independently or they may cover companies along with the senior analyst. As the senior analyst, you cover companies on your own. From there you can be promoted to different levels, depending on the firm, but your duties remain essentially the same.

It is not required that you have a scientific background to be a good analyst; in fact, many analysts are not Ph.D.'s or M.D.'s. However, I believe that a broad scientific background is very helpful. It is necessary to understand the drug development process and some scientists are never exposed to this area. Some business experience is also important. You need to be able to converse with business people in their language and to understand their problems. This type of experience can be obtained by getting an MBA and/or by working in the biotechnology or pharmaceutical industry.

The industry experience should also include some involvement in strategic planning or business development, even if it is peripheral to your scientific responsibilities. The goal is to gain some experience in how corporations operate. Scientists sometimes have a tendency not to see the broader picture because they are so involved with the details. It is also helpful to read about companies and the stock market to understand the world you are entering and how your clients, institutional investors, think.

A combination of skills is required to be a successful biotech analyst. These skills include written and verbal communication skills, scientific and medical knowledge, knowledge of the drug development process, the ability to deal effectively with a diverse group of people, and the ability to rapidly assess the potential of a corporate strategy.

Being a biotech analyst is an exciting and rewarding career. Looking for excellence in companies and watching—sometimes helping—them achieve their goals is gratifying. It is exhilerating to witness the development of whole new areas of medicine that have the potential to revolutionize they way we treat disease.

8

INVESTMENT BANKING:

Dreams and Reality

• • • • • • • • • • • • • • • • • •

Peter Drake, Ph.D.
Executive Vice President, Vector Securities

I have always loved science. As an adolescent, I enjoyed learning about aquatic biology and botany fishing in the solitary streams of southern Missouri. But the real fun began as a teenager when I got a summer job in 1967 as a complete grunt in the Surgical Research Laboratory at St. John's Mercy Medical Center in suburban St. Louis. I was 14 years old and had a job where I could wear that all-important white laboratory coat.

That coat was the only ego-boosting aspect of that first job, which tested my resolve to become a research assistant. Much of the experimental work in the lab involved surgical procedures on dogs, which meant I was assigned to the beagle patrol. Yes, I cleaned the dog pens daily in the sweltering heat of the St. Louis summer. I passed this first test, and the following summer I was rewarded with the air conditioning of the lab and the minimum wage ($2.50/hour).

After learning the rudiments of chemistry, biology, and physics in high school, in 1972 I headed off to Bowdoin College in Brunswick, Maine, intending to major in biology and later attend medical school. I was determined to choose a career path distinct from my father, who, as a farm boy with only a high school education, was one of the founding employees of

McDonnell Aircraft Corporation, known today as McDonnell Douglas (or, since the recent merger, Boeing).

But my dream of attending medical school was dealt a severe blow in my first semester of freshman year. Despite working harder than I ever had, my grades were straight Cs. I still loved science, but it seemed clear that my academic interests had to be applied to a career that didn't require the straight-A record needed to get into medical school. I changed to focus on the pursuit of science as an intellectual passion, all the while keeping an eye toward other career applications open to an undergraduate double major in Biology and Russian.

THE PLAN

By my senior year in 1976, it was clear that two areas of science—molecular biology and neuroscience—were poised to undergo a revolution. It was also clear to me that these scientific advances would lead to remarkable opportunities in industry. So my plan was to get a Ph.D. and position myself to capture new career opportunities in the industrial applications of biological sciences.

I chose to attend graduate school at Bryn Mawr College (with an all-female undergraduate school) in the fall of 1976 for two reasons: first, I could pursue a double degree in biochemistry and neurobiology with professors I truly admired and, second, after 8 years at an all-male prep school and 4 years at Bowdoin (where mine was the second class to admit women), I could instantly improve my odds of getting a date.

I spent my first 2 years in the advanced seminar program, which had a strong emphasis on independent study. My summers were spent at the Marine Biological Laboratory (MBL) in Woods Hole, Massachusetts, working on experiments to support my Ph.D. thesis.

It was in Woods Hole in the summer of 1978, while sitting on the beach reading the Sunday *New York Times*, that I learned about the first sign of a convergence of biology and industry. A patent covering an oil-eating bacteria had been granted to General Electric Co. This so-called Chakabarty patent was the first granted on a new life form. It was absolutely clear to me that if one could patent new biological discoveries, industrial applications were not far behind.

THE MENTOR

The harsh reality in 1980 was that I had a Ph.D., a fiancee, and a job in the Department of Anatomy and Developmental Biology at Case Western Reserve Medical School paying $13,380. By accepting this job, I allowed my career to move along a classic academic route in pursuit of my love for science. Unfortunately, I had achieved the "easy" part of my plan—the advanced degree—but I lacked the career-specific vision and the means to execute this vision.

I didn't know it at the time, but what I needed was a mentor. But within the span of a few months, I found two mentors. The first would provide me with the business vision to pursue and the second would give me the intellectual confidence to execute the plan. My business mentor took me back to St. Louis. Dr. Louis Fernandez was named chairman of the board of Monsanto in 1980, capping a distinguished career at the company where he began as a bench scientist. Monsanto was positioning itself for the future of agricultural biotechnology under Dr. Fernandez's leadership. He was also on the board of directors of Biogen, at the time a fledgling biotechnology startup in Cambridge, Massachusetts, led by the Nobel Prize laureate, Dr. Walter Gilbert.

In high school, I had often played tennis with Dr. Fernandez. So I took a risk and wrote him a letter describing my interest in combining a background in science with work in industry. Five days later, I received a return letter inviting me to a meeting in St. Louis. During our 2-hour discussion, we spoke about the coming revolution in the biotechnology industry and Monsanto's keen interest in capitalizing on the opportunity.

Dr. Fernandez told me to pursue my postdoc but concurrently to take some accounting and finance courses. He told me that Monsanto would have many opportunities for someone with my skill set, but if he were in my shoes, he would go to Wall Street and become an investment analyst. His view was that the biotechnology industry was going to emerge as an important new industry with significant capital requirements. Also, there were going to be many public biotechnology companies in need of analytical and scientific insight. Well there it was, presented on a silver platter—undeniably the best advice that I have ever received. I now had an executable plan.

My other mentor's influence was gradual in developing, but his teaching had a profound impact on the skills I draw upon as an analyst and investor. Dr. Raymond Lasek, a noted neuroscientist, taught me to work and think independently; to be curious and creative; to write succinctly and to communicate clearly with strong and persuasive selling skills; to take risks and to be scared; and above all, to challenge the wisdom of the status quo.

THE FIRST BREAK

During the three years I spent in Dr. Lasek's lab, I learned completely new fields of science, but most important, I learned that I could teach myself almost anything. My research project was highly interdisciplinary, focusing on neuroscience, cell biology and biochemistry. At night and on weekends, I followed the stock market, invested in health care stocks, and studied the ascent of the biotechnology industry.

I prepared a résumé, spent months refining my cover letter (the written description of what I wanted to do in a career), and interviewed for a variety of industrial and consulting positions. This was all intended as preparation for the trip to Wall Street. I learned about the Wall Street firms covering health care and biotechnology stocks; I studied the backgrounds of the biotechnology analysts; I talked to investment professionals to learn about the investment process.

From this effort, I knew that I had a deficiency, and curiously enough it was an academic deficiency. My scholastic background lacked an academic pedigree recognizable on Wall Street. Specifically, I had not attended an Ivy League school or a leading business school. Either or both of these represented an entrance ticket to a job on the Street. Further, my family lacked Wall Street contacts. But an M.B.A. seemed out of reach—I was 29, married for 3 years and still making less than $15,000. I was, however, hungry and resourceful.

I then learned about a summer program sponsored by the Wharton Business School at the University of Pennsylvania, designed to train Ph.D.'s in business fundamentals. Instinctively, I knew the potential importance of this program. My summer in Philadelphia at Wharton was fantastic—tremendous teachers, dedicated administrators, and a diverse and an enthusiastic group of colleagues. The time was right to put this all to work.

I interviewed with three Wall Street firms and I chose to join a distinguished research department at Kidder, Peabody & Company as a biotechnology analyst. My group leader and Wall Street mentor, Arnie Snider, later told me that my cover letter and résumé stood out among the many people Kidder was interviewing. The refinement of these two documents over the previous year had paid off! In addition, I interviewed well, I was knowledgeable about the securities business and the other Wall Street firms, I had a view of biotech stocks and an understanding of the biotech industry.

After I was hired, my boss, George Boyce, showed me my office, gave me free reign in defining the stocks I would follow, and left me alone for a year as I traveled throughout the United States visiting companies and attending scientific meetings. I made $85,000 my first year in the business. I was hooked!

THE SKILL SET TO BE AN ANALYST

Having hired and trained scores of analysts over the past 15 years, I would say that there is no single trait or skill set necessary for success. In fact, I will admit that being the oldest biotech analyst on the Street does not mean that I have all the answers or that the learning process is over.

Before describing the various important characteristics of being an effective analyst, I'll start with the critically important things I've learned. Arnie Snider taught me that there is a difference between companies and stocks. Biotech companies are about people, technology platforms, business and financial strategies, and products; stocks are pieces of paper. An analyst's primary job is to recommend stocks that go up in value and to avoid stocks that go down in value.

It is also critical to recognize that in the investment business, we live in a relative world as opposed to an absolute world. By this, I mean that institutional investors are supremely performance-oriented and they are graded on the basis of the performance of their funds relative to the market's overall performance. If an analyst's recommendation is up 25% when the S&P 500 has appreciated 30%, that means the stock has under performed the market. I teach our analysts and investment bankers that we are actually in the paper business—we analyze it, we trade it, we issue it. Consultants analyze companies and strategies, but securities analysts are stock pickers.

The next step in the process is communication—both verbal and written. When an analyst has found a stock that he or she believes is undervalued (the stock price does not reflect the company's value), the key is to market the idea to both the firm's salesmen and directly to institutional clients. If an investment idea can't be packaged and communicated succinctly, the idea will not rise above the competitive noise level of other analysts at other investment firms pitching their ideas.

Beyond these big-picture issues, the skill set for a biotech analyst includes honesty, professionalism, a strong work ethic, a competitive spirit, being inquisitive and skeptical, having a solid accounting and finance background (i.e., an M.B.A.), having a strong knowledge of science and technology, paying attention to detail, and having strong financial modeling skills.

A TYPICAL DAY

The brokerage business begins with the morning research meeting. Analysts stand in front of the salespeople and traders and make a 2- to 5-minute sales pitch focusing on the reasons to buy or sell a particular stock. Then, salespeople and analysts hit the phones, marketing the idea to institutional clients on the "buy side"—specifically, portfolio managers and analysts.

Nearly every day analysts speak with companies, either in phone conferences with the top management contacts or in one-on-one meetings with these executives in the office or during site visits at company headquarters. These contacts with company managements become the source of future comments for the morning meeting. News items by companies that emerge during the day are analyzed and comments to the sales force are directed to institutional clients. Finally, considerable time is spent in writing formal research reports.

A typical day spent out of the office involves one of three activities: site visits at companies to meet with key senior management; attending industry conferences and/or scientific meetings; and marketing to institutional clients in group meetings (e.g., at breakfast, lunch and/or dinner) or in one-on-one meetings. Herein lies, in my opinion, the most enjoyable (and potentially the most stressful) part of the job.

As I write, I am on a marketing trip. I flew to San Diego last night and had a group dinner (which ended at 11:00 P.M.); I was on the morning research call at 5:00 A.M. (Pacific time); after three one-on-one meetings in San Diego, (1-hour meetings), my West Coast sales representative and I drove to Los Angeles. We had three one-on-one meetings, one of them with a client who called me stupid for not having upgraded my recommendation on a stock at 10 that I had lowered my rating on 4 weeks ago at 16. (The stock was up 2 points today to 14.)

I currently am on a flight to Portland, Oregon, for a dinner meeting and two one-on-one meetings tomorrow morning; then I will catch a flight to Seattle for a group lunch followed by three one-on-one meetings and a late night flight (with a tight connection through Denver) to arrive in Chicago (our home base) at 1:00 A.M. Then, the morning call, two conference calls, and a day off (maybe) on the weekend. So, an analyst must possess endurance and a thick skin.

From the financial reward perspective, it is not uncommon for industry-specific analysts (biotech, medical devices, health care services, and so forth) to start with a combination salary/bonus of $100,000 to $150,000. From there, it's a matter of how much revenue you generate for the firm. In one way or another, analysts and bankers are rewarded based on the money they bring into the firm, which can mean some pretty big bucks. A good stock picker has the opportunity to double or triple her initial salary, based on time at the firm and performance.

WHERE I HOPE TO GO FROM HERE

I joined Kidder, Peabody & Co. in October 1983; I loved the firm and I made lasting friendships with colleagues and clients. I was honored by becoming a partner within 2 years and achieving the Number 1 ranking on the Institutional Investor All-Star Team.

I cofounded Vector Securities International, Inc., in 1988 with two remarkable partners, colleagues, and friends—Ted Berghorst, an investment banker, and Jim Foght, Ph.D., a former pharmaceutical executive turned investment banker. Together we have formed the Later Stage Equity Funds I and II with a total of $230 million under management, designed for private negotiated investments in life sciences companies. In addition, we co-founded Deerfield Partners, a $400 million health care hedge fund run by Arnie Snider in New York City. Vector Securities International has been involved in more than 200 transactions and has raised roughly $3 billion in capital for the life sciences industry.

I spent the first 6 years being an analyst and running the research department as director of research. This worked as long as Vector was a small organization. As the company grew, the research department expanded to 20 professionals. I had to decide if I wanted to oversee the management of the research department or go back full-time as an analyst. By the end of 1994, I chose to be an analyst and we hired the person who had run research at Drexel and was CEO of County NatWest Securities.

My current responsibilities as executive vice president and biotech analyst include participating on the operating committee and the board of directors. The operating committee meets weekly for a couple of hours to oversee and review the firm's operations. I am also a partner in the Later-Stage Equity Fund. We just closed the second fund in this family with $180 million, and Genentech as the lead special partner. I had followed Genentech for years as an analyst; I no longer cover them as an analyst, but they are showing their faith in Vector by investing a substantial amount. We spent the last quarter of 1993, all of 1994, and part of 1995 raising $51 million for Fund I, providing investors with a performance that is up 90%. This paid off—it took us only 4 months to raise $180 million for Fund II.

All of this sounds great. But I am a biotech analyst; I am only as good as my last recommendation. I am competing with biotech analysts at other Wall Street firms for the attention and brokerage business on the buyside; I am competing for the time and attention of our sales force. My stock picking over the past 12 months has included inconsistent big winners and several losers. Year-to-date, the biotech stocks are off 3% while the S&P 500 has appreciated 23%. Our salespeople are angry with me and our clients are dissatisfied with my performance.

The brokerage business is a tough game. There is a reason that I am the oldest biotech analyst (chronologically 43) and I have survived the longest—14 years. I still go to sleep tired every night and wake up scared every morning. But I live in and thrive on the brokerage business. I am an optimist, looking for important drugs that may cure diseases and create untold wealth for investors. I am also an opportunist, willing to change my investment rating on a stock when my experience and judgment dictate. I will retire as an analyst someday.

chapter

9

BUSINESS DEVELOPMENT:

Making Deals with Science

Ron Pepin, Ph.D.
External Science and Technology, Bristol-Myers Squibb

MY EVOLUTION

I didn't plan to leave the lab and become a dealmaker. It just sort of happened. After busting my hump at lab benches next to autoclaves, after labeling an infinite number of Eppendorf tubes for endless sequencing reactions, after a variety of battles with ^{32}P, after extraordinary moments of joy with successful experiments and new hope after failures, I never planned to throw in the lab coat and its associated rewards for a tie. It just happened. Suddenly, I was dealing in the millions of dollars instead of plaques, telling people about the merits and flaws of their research from a commercial instead of a scientific perspective, and wading through buckets of biotech hype. And, to quote Maxwell Smart, loving it!

As a Ph.D. molecular biologist, I had been at the bench for a couple of biotech companies before I decided that I had been here and done that—once you've cloned a few genes the excitement can wear off. Especially when they now make kits for all the cutting-edge work I used to perform! Before I traded in my pipette man for a Mont Blanc pen, I had no idea what was involved in business development.

Even as a researcher at a small biotech company where everyone knew each other and their functions quite well, the business development guys were just part of the typical biotech "dog and pony show" that was constantly staged to keep the company afloat and me with a paycheck. Quite honestly, as with most jobs, you really don't know what it's all about until you do it. I was pleasantly surprised to find my true calling.

What Is Business Development?

Business development involves bringing "things" from the outside into your company in order to enhance your company's business. The "things" can be intellectual property (e.g., patents or patent applications), technologies, know-how, compounds, drugs, research collaborations, equity investments in a company, and so on. This variety of items is what makes the job so challenging and exciting. The acquisition of these "things" can be done on an exclusive (for your company only) or nonexclusive (shared by two or more parties) basis. Financially, the transactions to obtain these "things" may involve the simple exchange of money for goods, property, ideas, or other creative means that ensure that both parties obtain what they want and need.

At Bristol-Myers Squibb, business development has been divided among several different groups, each of which has its own niche. This also allows each group to have particular expertise among its members. As a part of the research division, the External Science and Technology (EST) group is responsible for the acquisition of early-stage technology from biotech companies, universities, and government agencies on a worldwide basis. The deals that this group performs range from simple licenses to drug discovery targets, receptors, or other forms of intellectual property, up to extensive research collaborations with biotech companies, which typically include up-front payments, licensing fees, research funding, payments for the achievement of clinical milestones, and royalty payments. The deals may also include an equity investment in the stock of the biotech company, which involves the valuation of the company and its platform technologies.

To complement the work of EST, a second business development group exists for licensing compounds that are in clinical development. The agreements put together by this group are similar in structure to EST deals (i.e., up-front licensing fees, research support in some cases, clinical milestone payments and royalties on product sales). Some deals also involve an equity investment. The client group of this licensing team, however, is broader than just the drug discovery division of research. Much more attention is paid to financial projections. Because the compound and its potential indication use have been identified, the market projections for such a drug are more predictable than something in early preclinical research as a part of one of EST's research alliances. The licensing group will also interact more extensively with the clinical division, which has a justifiably heavy influence on the licensing deal. Marketing considerations are also more important for these later-stage deals.

The mission of EST is to enhance the ability of the research group to discover new drug candidates. The job involves both the proactive and reactive evaluation of technology in order to match its fit with the needs of the research division of BMS. On the proactive front, EST visits universities, government agencies, and biotech companies to learn the novel technologies that they have uncovered and, if available, to evaluate its value to BMS. Alternative proactive approaches include attending scientific meetings, partnering conferences, and investment banker meetings to track down the sources that are discovering cutting edge technology. Such "one-stop mall shopping" is a time- and cost-effective way to survey multiple opportunities in a short period of time and to build one's mental database of the playing field in the business.

In addition to the excitement of proactive prospecting for appropriate opportunities, EST and other business development professionals must also react to unsolicited proposals. We receive on the order of 400 to 500 proposals per year from any university technology transfer agent or biotech business development representative worth their salt. EST personnel serve as a first screen of such proposals. It is our responsibility to filter out and reject the real turkeys (I have received nonconfidential data on alien spaceship warp drives, no kidding!) and to discuss with the appropriate scientists within our organization the technologies that have substance and merit and that may fit our research strategy.

Once a technology has been identified as one in which BMS indeed has an interest, we quickly start to move matters ahead. If the initial disclosure about the technology was nonconfidential in nature (i.e., a very brief and usually rather uninformative document designed to whet our appetite for the technology), we will request to see confidential information under the rules and regulations of a Confidential Disclosure Agreement, or CDA. A CDA is a legal document designed to protect the disclosing party from having their confidential intellectual property (such as a novel cloned gene) from being ripped off by the people who view the documents.

If the confidential data package (with an emphasis on the existence of data) still looks impressive, BMS usually invites the scientists who made the discovery to make a scientific presentation to our in-house scientists. The business development person attends such sessions in order to learn about the proposed technology so that he or she can help value the information for BMS and formulate a deal strategy to bring the technology into BMS.

At this point, a commitment needs to be made by both sides. If there is interest from both parties to move ahead and strike a deal, then the proper structure for the interaction must be formulated. Both sides must also negotiate in good faith such that, if a "win–win" agreement can be reached, both parties will execute it. I have been involved in negotiating major proposed alliances where BMS has moved forward in good faith, only to have the other party decline on doing the deal at the eleventh hour. One actual deal was closed to 11:59 P.M. when I had a huge equity investment and R&D alliance being voted on by the BMS board of directors, only to find out that the other party had been negotiating with several parties at the same time!

I have also worked with skilled professionals who know their job and the value of their technology. Such individuals make the negotiation process as painless as possible. Most people would assume that I would rather work with a rookie negotiator of whom I could take advantage. In reality, it is much more difficult to negotiate with someone who has no knowledge of what is included in a typical contract and what terms are the norm for the industry for a given type of deal and technology. It's about the same as teaching your surgeon how to operate during your heart transplant. It just seems to go more smoothly if they have been there before!

In any event, the most exciting component of the deal revolves around the negotiation. To those who have not been there, it would appear that, once the horse trading over money is completed, the negotiations are over. But, nooooooo! God invented lawyers! That means that the good old English language has to be put into "legalese" before it becomes a final, unambiguous document. And getting there, if you ever do, is the most laborious, time-consuming part of the whole process.

In all seriousness, the lawyers play an important and helpful role for both negotiating parties by crafting language that will protect their respective clients. It's just that the dotting of the i's and crossing of the t's can be a painful experience. Each party usually has its own sensitivities that need to be recognized and met by the other negotiating party. And some sensitivities are more sensitive than others!

You have to pick your battles and go to the mat on the most important issues, while the ability and need to compromise must arise on some other lesser issues to make the negotiation work. Holding your ground on each and every point will only serve to kill the deal. Even as the party with the "upper hand" (i.e., we have the money), you simply cannot insist on your way on every issue if you want to complete a deal that is favorable to both sides. My personal philosophy is that if both parties leave the negotiation table feeling good but with some bad taste in their mouth, the negotiation was a success.

SOME OF THE CHALLENGES

Often the external negotiations (i.e., those with the third party with whom you are doing the deal) on any alliance are easier than the internal negotiation process that must take place within your own company to sell the deal. The internal selling of a deal involves the buy-in of many diverse groups of people beyond the original scientists who became excited by the deal—finance analyzers, legal, the tax department, other business development groups, not to mention management at all levels. Because these groups have different backgrounds and are not commonly trained as scientists, the ability to translate complex scientific concepts into lay language becomes an important skill to possess for this job. I view it as the innate ability to teach. Just remember back to graduate school where the hot shot research professors were generally the very worst at explaining and spreading their knowledge to students.

If you want to sell your deal, you better know not only the scientific complexities but also ways to convey the financial importance to your organization's management! Use graphs, figures, analogies, anything that will convince those who control the purse strings that the deal is truly important for your company. This also is true for the small companies and universities trying to sell the technology. Their internal sell job is somewhat different since it's generally easier to sell someone on accepting money rather than on spending it.

However, since most biotech and university managers are trained scientists rather than business people, you probably have to provide similar data on the fairness of the deal, benchmark it against other similar deals, and demonstrate the value to your organization in order to make it fly. The financial gurus will provide you with estimates of the market value of the products that may arise from the alliance that will be an important concept to relay to your business-focused management. And, of course, they will want to see that the legal and tax divisions have given the contract a seal of approval.

If an equity investment is involved, a detailed explanation of the price per share that you are paying to a biotech company will be required. Here is one area where experience in the industry will really pay off. It's easy to value a publicly traded biotech company, but producing a valuation for a private company is an art form. It involves evaluation of their " platform technologies," comparisons to comparables, and their previous financing.

By factoring in all these variables and the level of interest your company has in doing such an investment, you can arrive at a price that is acceptable to both parties. Your management may, however, take some comfort in having your valuation backed up by an external evaluation firm. In any case, you can always protect your investment against any pumped up price sought by the biotech company by instituting antidilution provisions in the equity document. These provisions allow you to make an investment at a set price but they guarantee that the company in which you invest will not sell additional shares to another party at a price lower than that paid by your company. If they do sell off cheaper shares, the antidilution provisions spell out how the company will compensate you for that transaction (e.g., they can give you additional shares to make you whole on your investment such that you pay the same lower price paid by someone else).

WHAT IT TAKES TO MAKE IT IN BUSINESS DEVELOPMENT

A typical day in business development involves a great deal of communication. Large phone bills and long hours at the keyboard answering e-mail are par for the course. Given that you may spend anywhere from 10 to 75% of your time traveling, the business development executive commonly may be found doing the above-mentioned communication on the road!

If oral and written communication skills are not your strength, stay away from business development! And shy introverts need not apply. You'll be eaten alive at the negotiation table if you don't like to speak up and counteract your counterpart sitting across from you. The art of negotiation can take several different forms, ranging from bluffing to putting all your cards on the table. The key in my mind is to be honest and fair, and to negotiate in good faith, with true interest in striking a deal.

One thing that initially attracted me to the business development field was the unique opportunity it affords a scientist to make use of his/her hard-earned degree in the sciences. In addition, it allows you to play a part in the business aspects of your company or university. My experience (bias?) has taught me that it is easier for a Ph.D.-trained scientist to pick up the business aspects of this job than it is for an M.B.A. type to pick up on the subtleties of the science. After all, we all have to balance our checkbooks and negotiate for cars. On the other hand, how many of your nonscientist friends have ever done a Northern Blot? Analytical skills are needed to fairly, quickly, and accurately evaluate a vast variety of scientific proposals, to uncover new research opportunities, and to analyze the business aspects of a deal.

I feel privileged to have seen the biotechnology field from both sides of the coin: from the lab bench of a biotech company to the wheeling and dealing as part of business development at a large pharmaceutical company. I believe that this past gives me the perspective to see how different aspects of a deal are important to different parties and how it is vital to satisfy the major needs of both parties. It has added to my ability to be creative in structuring alliances that are workable for both sides of the agreement.

How Do You Get Here?

Getting into this racket is not necessarily easy. Most large companies and biotech firms justifiably require that you have experience in order to hire you. That said, one way to join the ranks of business development is to work at the lab bench within a company so that you develop a strong positive reputation. Such a " known quantity" can parlay that reputation to gain the confidence of company management and to secure a business development position.

I would rather train a competent scientist from within my company who possesses other desirable personal and professional characteristics than to hire an individual form the outside about whom I know very little (how much do you really learn in a one-hour interview?). This allows me to mold them in such a way that they perform their business development duties in a manner and style of which I approve.

I also have a strong bias toward making these choice opportunities available first to people who work within my company. Given the small size of this function within most companies (large pharma companies like Bristol-Myers Squibb employ about from 4 to 5 people in EST and another 6 to 7 in licensing, whereas most small biotech companies have no more than 2 people in similar functions), you must choose and develop your team very carefully in order to be effective!

An alternate mechanism for joining the business development club is to pay your dues at a university technology transfer office rather than working your way up via the bench. Most major universities have an office that employs individuals who are responsible for the out-licensing of university technology to large and small companies. Because universities generally pay less money and the deals are most often less complicated in structure than biotech deals with big pharma, this is a prime training ground for learning this trade.

A few years of experience in this environment gives you the training to make the leap to an industrial position. Once you have completed some deals at a biotech or pharma company, you join the elite few business development people who actually have experience in the field. At that point, you can write your own ticket. You will find that the headhunters, the best source of such positions outside of word of mouth, will be calling you on a regular basis! Salaries for novices in the industry fall in the range of $50,000 to 75,000 (again, universities pay less) and seasoned professionals can make far in excess of six figures.

In line with the hype involved in the biotech industry, salary is not necessarily correlated with one's experience or abilities in this field. I have taught Business Development 101 to many a biotech novice/idiot who made more money than I did (Gee, who does that make the idiot?). I once had a rule that I would not do a deal with a biotech company whose business development person drove a better car than I did (I was driving a 4-year-old Ford Taurus at the time). Rather than giving up doing deals at all, I have since relaxed this rule.

In conclusion, a position in business development gives you the opportunity to use your scientific training and creativity without being anchored to the lab bench. It provides the opportunity to travel, to attend scientific and business conferences, to learn about the business aspects of the pharmaceutical industry, to explore scientific areas well outside of those that your research focused you on, and to make new acquaintances in one of the most fascinating industries available to you as a scientist. Although it may on a few rare occasions frustrate the hell out of you, the rewards far exceed the negatives. Now go out there and do a deal!

10

•••••••••

ENTREPRENEUR AND COMPANY FOUNDER:

Starting Your Own Company and Surviving

•••••••••••••••••

Ron Cohen, M.D.
CEO and President, Acorda Therapeutics

Although I started out in a medical career, for the past few years, I have been creating and building a new biopharmaceutical company, Acorda Therapeutics. The aim of the company is to create novel therapies to treat spinal cord injuries (SCI) and potentially other neurological damage. Prior to founding Acorda, I was part of the start-up team for Advanced Tissue Sciences, Inc., of San Diego, California. The work of building and growing new companies is hard, but the rewards are great.

My personal goal is to build a company that can develop restorative therapies for spinal cord injury and other neurological damage; and to build a model of a company that works efficiently, and mobilizes as much of the talent out there as possible to work toward a common goal.

HOW DID I GET HERE?

During my internal medicine residency at the University of Virginia, I saw my first spinal cord injury (SCI) patient. That experience stayed with me even after I went back to New York City and practiced medicine while pursuing the other love of my life—the theater. For a while, I lived a dual life as medical director of a clinic in the Wall Street area and as an ER doc, while also taking acting classes, auditioning and working in commercials, off-off Broadway plays, and as a contestant on *Jeopardy*. I was having a wonderful time, living with one foot in each of two very different worlds.

But then in May 1986 friends from my medical school class at Columbia called and told me about a scientist friend of theirs, Dr. Gail Naughton, who was starting a company around her work in tissue proliferation. She needed an M.D. to help build the company, and who was good at making presentations. An "acting" M.D. seemed like the perfect solution. My friends had shown Dr. Naughton tapes of me being interviewed for the local news and as a contestant on *Jeopardy*, and she decided that she had found what she was looking for.

I had never considered going into the business world—but this made no difference to Gail. At my first meeting with her, her husband, and her two kids, she pulled out a piece of paper with three lines of writing that essentially said, " I, Ron Cohen, will work for Marrow-tech for $X," and she said, "Sign here!" While I wanted to think about it, Gail said, "You don't understand. I want you to sign here!" So I did, and that is how I got into biotech—it was an impulse based on the way Gail and I clicked; it just seemed to be the right thing to do.

This simple exchange of "sign here" provided me with the most profound change in my life. I plunged in without knowing anything about business, and neither did Gail. A dentist friend of a relative put in the initial money, which ran out after a couple of months. But I was hooked—I worked for no salary and worked part-time in the ER to support myself while Gail was working in the lab at Queensboro Community College to make ends meet.

Two years later, we took the company public, recruited a very experienced pharmaceutical executive, Art Benvenuto, as CEO, and grew Marrow-tech in Advanced Tissue Sciences, Inc., which is now a major biopharmaceutical firm with approved products entering the marketplace.

A DAY IN THE LIFE OF A CEO

There is no typical day when you head up an early-stage company. Since I am the only full-time employee other than my assistant, I do anything and everything. The responsibilities are overwhelming at times. At this stage in the company's development raising money is key, along with keeping the "virtual" organization motivated and on track to meet goals, attracting new members of the scientific and business advisory groups who can help us grow and take things to the next level, strategic and financial planning, accessing new technology for Acorda to develop, and maintaining an intellectual property portfolio. We have to generate and maintain technology licensing agreements, patent applications, research agreements, and corporate partnerships. My assistant, Tierney, now does the in-house financial management and office management, including responding to patients and editing video PR pieces.

In addition to dealing with company staff, a CEO also must work with the company's board of directors. This group should provide experience and expertise apart from the in-house management team's, they should provide access to networks of potential sources of funding and corporate alliances, and they should critically evaluate the decisions and progress being made by the company. Some of the board members represent groups that have invested heavily in the company.

This means that you have a diverse group of people on the board, each with a different agenda and different past experiences. Managing the board can be a challenge, but a strong, well-balanced board can help a CEO stay ahead of changes in the business and financial environment and it can allow the CEO to steer the company through rough waters.

The general timetable for each day is about the same, but the activities can be very different. I usually get up at 6:30 A.M. and read the *Wall Street Journal* and the *New York Times*. I head for the office around 8 A.M., where I read e-mail and make calls. The rest of the day can be taken up with traveling to meetings with potential investors, potential or existing corporate partners, scientists whom we wish to recruit, or to industry meetings to make contacts. If I am not traveling, the day can include anything from chairing a scientific advisory board (SAB) meeting, to conducting a conference call or a meeting to plan our clinical trial program. Each day, I discuss ongoing projects with Acorda's chief scientific officer based in New York and our research director in North Carolina. The day usually ends between 8 and 10 P.M., with the occasional allnighter. I work one or two full days on the weekend, but I try to keep at least a half-day for fun.

Here is what I did last week: On Monday, I wrote an ATP grant to apply for $2 million to fund a program, I had telephone conference on clinical trials, and I worked on financial pro formas with our acting chief financial officer. On Tuesday, I flew to an industry meeting in Boston to make a presentation on being a virtual biotech company, and I had dinner with our chief scientist and our corporate collaborators to discuss the clinical trial program. On Wednesday, I attended the rest of the conference and met with a company consultant to discuss corporate development strategies. On Thursday, I was back in the office and met with our chief scientist and a potential pharmaceutical partner from Europe, and then met with auditors to get ready for an upcoming audit to prepare for our next financing.

On Friday, I met with a potential candidate for full-time CFO and vice president of business development and I had a working session with the acting CFO; I worked on a revision of the executive summary and business plan in preparation for the financing; I went to the chief scientist's lab to discuss recent data; and I worked on a slide show for the financing.

For the next 5 to 10 years, I plan to stay with Acorda and get it to the point where I feel I've accomplished my key goal—successfully developing breakthrough products for restoring neurological functions, which is one of the big remaining frontiers in medicine.

WHAT IT TAKES TO RUN THE SHOW

Growing a company from the ground up is definitely "seat of the pants" learning—the first 5 years at Marrow-tech were exhilarating and intense, with a series of unremitting crises on a daily basis that threatened the very survival of the company. This setting is exciting, it requires high adrenaline levels, and it is in certain ways like being at war with buddies. The most exhilerating point came when we lifted our heads out of the trenches and saw that we were actually making progress and building a company.

I became a confirmed adrenaline junkie—which is a good thing. Stress becomes a constant part of your life. Once you take other people's money, and are responsible for watching over their investment, everything changes. Even more importantly, once you have employees who look to you for their paycheck and health benefits, you realize that your actions impact a wide group of people who depend on you to keep things running. There will always be some crisis or other to be resolved. I suggest taking up boxing or racquetball!

One of the most important skills to have for anyone who is involved in managing a company is the ability to get other people excited and engaged. This is particularly crucial in startup companies, where a very small team of people works very hard in a very high-risk situation. You have to keep people willing to work long hours for what is frequently low pay. At any given point, the company can appear on the threshold of financial or scientific disaster, and the executives at the top have to keep everything moving forward in spite of this. You must be able to motivate folks to take on a broad range of tasks, since a young company can't afford to hire enough employees to have specialists.

The CEO had better know exactly what they want to do with the company, what the goals are, and then he or she must be able to communicate with others who must come together to make this goal a reality. Lots of people start companies with the vague idea that they want to be in a certain area of research or clinical development without really having thought through just how to turn that idea into a sustainable business.

You must be able to pursuade a group of people from different walks of life—scientists, technology transfer folks at universities, other companies, investors of various types, physicians, lawyers, journalists—that your business is worthwhile. You must enlist their support of your cause.

Communication is the second great task. If you personally are not comfortable communicating in public, you must bring in someone to be the public voice. The CEO doesn't have to be charismatic (although this certainly helps), but you must be able to communicate clearly what you are doing, why it's important, and you must get the outside world excited about it. The company lives or dies based on your ability to communicate—if outsiders aren't supportive, you will run out of money and the company will die before accomplishing anything worthwhile.

You must be able to decide not only what the company's goals are, but also how you will implement your decisions and stay on track through the inevitable crises. People (employees as well as investors) want to know that the company's leadership is a very clear beacon keeping them off the rocks. You must stay the course, and keep people optimistic rather than panicky. As Woody Allen says, 90% of success is just showing up. It takes about a decade for a new pharmaceutical to move from discovery into the marketplace. The company must survive long enough for an opportunity to come along that you can exploit into a business opportunity. You must be an optimist in order to keep others fired up. If you are pessimistic, if you constantly look to the possible negative outcomes, why should others persevere?

You need to have the ability to collect and synthesize data, the ability to interview people productively, and to ask the right question at the right time to get the information you need. An important part of my training as an internist was learning how to get a good medical history. A CEO must have the ability to identify people to bring into the company who have the right talent and personality traits, and the CEO must be able to find the right technology to in-license, the right sources of financing, and so forth.

I have found that the ability to identify early on the people who will help you and those who will hurt you is a critical, but apparently rare, skill. Young companies are constantly approached by individuals who portray themselves as "friends and supporters" of the company. But often you will find that this friendship lasts only long enough for them to extract whatever they need from you. Sometimes you need to do business with them anyway, but it is always better to know when you are dealing with a real friend and when you are not.

Learn to trust your gut reactions to people and situations, as opposed to overintellectualizing everything. In my experience, in a fight between my gut and my head, my gut is usually right. For example, we have pulled to-gether a scientific advisory board of leading scientists working in different areas related to SCI. Prior to working with Acorda, these folks often com-peted with each other for grants and publications. As part of our SAB, they work as a team, sharing information and ideas, as well as technology. This requires a strong team spirit and a willingness to work together without be-traying the confidence of the group.

My chief scientific officer and I met with a scientist whom we liked im-mensely and who we thought could provide crucial technical expertise to the SAB. This person elicited a very strong reaction from other scientists in their technical field. Numerous people in the field gave us very negative feedback—it was a real dilemma. But my gut said strongly this person was important for achieving our ultimate goal. I went with my gut. Three years later, this has shown itself to be one of my best decisions. This person is one of the key performers on the team, and a real asset to the others on the team. Had we gone with our heads, we would have made the wrong decision.

This gut instinct will be important when making decisions about poten-tial corporate partners, bankers, and employees. Also, young companies tend to use consultants to help handle some of the more specialized tasks, such as business development, financial analysis of deal points, corporate communications with the press and the financial community, clinical and regulatory affairs, and manufacturing. Because these folks are working out-side of the company, you need to identify those folks who really have your best interests at heart and who are conscientious and honest.

THE PROS AND CONS OF BEING THE BOSS

What I like the most: I enjoy building teams of extraordinary individuals to do extraordinary things. It is a rush to get a group of very talented and mo-tivated people of good will to come together to do something challenging and worthy. And we are all working toward goals that we can be very proud of in a humane sense. This is the best of all worlds.

What I like the least: Invariably, our progress in the science exceeds our ability to fund the work. I am impatient and hate the time needed to raise money. Biotech firms never seem to have enough funds to implement fully their strategic plan, to build the team as fast as they want. This means that I have to spend an inordinate amount of time gathering the capital until the company is able to fund itself. This is particularly true in biotech, be-cause the product development path is long (7 to 10 years), and it takes years to get product to market and generate significant revenues.

One of the problems with running a lean machine is that the money limits how many employees the company has and what they are paid for working in a very high-risk environment. I didn't receive a salary for the first 3 years. Instead, I worked for a significant chunk of founders' stock, which will some day be very valuable if we make Acorda a success. Recently, the board voted me an annual salary of $120,000, based on the closing of our first corporate deal and the company having made sufficient progress and having raised sufficient funding.

HOW CAN YOU GET THIS JOB?

These positions are not advertised. You generate the opportunity yourself by starting a company, or you find someone who has just started a company and join their team to learn the process (as I did with Advanced Tissue Sciences).

I founded Acorda from scratch, and on my own dime. This meant several years of deciding which area I wanted the company to pursue, plowing through the technical literature and attending scientific and clinical conferences to find out the current status of SCI research and therapeutic development, identifying the top scientists in the field and pursuading them to work with me, figuring out what the other biopharmaceutical companies were doing in the area, getting the various not-for-profit foundations that fund SCI research to cooperate rather than compete. Then I had to write a business plan that made scientific, clinical, commercial, and financial sense. Hardest of all, I had to round up sufficient investment dollars to fund the company. This last task never ends, but you won't have a company without the first infusion of dollars.

If you want to become involved with someone else's company, you have to use your network to reach them. Start with people you know who are already in the industry or are leading scientists in the field. For example, the best way to find out about opportunities in a startup working on new drugs to treat Alzheimer's disease is to identify and network with the scientists who are developing key technology in that field.

Keep your ears open—if there is a talk being given by someone from a company, go and meet that person and network—this is a critical skill if you want to get into business. If any of your friends or professors have contacts in the relevant arena, make sure you network with them and express enthusiasm and interest.

What about entrepreneurs who come right out of academia? In a strange way, ignorance of the business world can be useful for entrepreneurs. One of the hardest things anyone could ever do is to start your own company. Had I known just how hard it really would be to get Advanced Tissue Sciences up and rolling, I might have been scared off. The advantage is to be naive—we were too ignorant to know we couldn't do this. Once we were committed, we just struggled to keep our heads above water. It was the hardest, yet most exciting, thing I had ever done.

However, at some point experience is crucial. My Acorda experience is very challenging—huge amounts of work and stress, but at least I knew what was coming. I strongly urge budding entrepreneurs to ally themselves with people who do have that experience, to balance out their ignorance. Scientists need to align themselves with people with management and people skills, who have been through it before. This requires swallowing your ego. It's hard to share control with others, but often it is a fatal mistake for entrepreneurs to try and go it alone. Strong scientific skills do not translate directly into excellence in other areas. A Nobel laureate does not necessarily make a good CEO.

11

• • • • • • • • •

CONSULTANT TO THE STARS:

Advising CEOs for Fun and Profit

• • • • • • • • • • • • • • •

Carol Hall, Ph.D.
Principal, BioVenture Consultants

Even though I had taken a classical academic science path—Ph.D. at Stan
ford in molecular biology, postdoc at DNAX Research Institute in Palo Alto,
California—my scientific career took a sharp turn away from the lab soon
after completing my postdoc. I knew the time to leave the lab had come
when I hated even the thought of pouring one more sequencing gel. The
thrill was gone. Too many days in the cold room had left me permanently
chilled toward the idea of countless more years of the same.

But what to do instead? I loved the intellectual aspects of science, and
was facinated by the idea of creating new therapeutics to help patients.
How could I promote that activity without having to personally run gels?

Keep in mind that what followed was not a well-planned career path,
but rather an evolving situation with random exposure to different pressures and opportunities acting as the Darwinian forces.

MY PATH INTO FINANCE

During my graduate studies, I took advantage of the close proximity of the
Stanford Business School to attend a few courses, just to see what that

world was all about. At that time, the late 1970s, the biotech industry was virtually nonexistant. This meant that most of the real job opportunities were in large pharmaceutical companies, a setting that didn't appeal to me particularly. While I didn't want to stay in academia, I enjoyed the informal environment that allowed easy collaboration and access to a broad range of disciplines.

I decided to do my postdoc at DNAX specifically because it was a small biotech company that would give me the chance to explore a move from the lab into the business side of the company. Unfortunately, I was there only one month before the company was acquired by Schering-Plough Corporation, and all the management was taken over by SP executives. It looked as though I was stuck back in the lab.

My big break came when my husband-to-be, another scientist, saw an ad in *Money Magazine* (*Science* isn't the only place to find interesting leads!) for a summer program offered by Wharton Business School at the University of Pennsylvania, to train Ph.D.'s in the art of finding a job in the business sector. I took a leave of absence from my postdoc and headed for steamy Philadelphia.

Those 2 months were invaluable. I learned to speak a new language ("businessese") and how to pick up the phone and cold call for interviews. The course highlighted the first year of business school—an overview of economics, finance, marketing, and management. The finance component was the most interesting to me, because it's very analytical and numbers-oriented. A large part of the focus on finance was the role of Wall Street—this was the place to be if you wanted to be involved with raising money for the biopharmaceutical industry.

Unfortunately, these short programs, once offered by several leading business schools, aren't around any more. A popular alternative is 2 years of business school, and I recommend that all aspiring consultants at least look at what these schools have to offer.

The most important skill that I acquired was how to pick up the phone and start looking for a job that wasn't listed in *Science*. After my crash business course, I quickly landed a job as a biotech industry analyst at a local San Francisco brokerage firm. Over the next several years, as I moved from analyst to investment banker, I picked up many business skills that would later serve as part of the basis of our consulting business. These included the ability to analyze a scientific organization from the perspective of the business and investment opportunity it offered, and the challenges this organization faced in raising enough money to realize that opportunity. I also gained insight into how Wall Street looks at science as opposed to how scientists look at it—very different views of the same world.

In addition, I took a 3-year self-study course to earn my Chartered Financial Analyst designation. This CFA substituted for an MBA and gave me more creditability in the financial community. The CFA is a credential that is well-recognized among the buy-side and investment professionals—it means you have a minimum level of proficiency in certain key areas related to investment analysis. I loved being in the financial community. These jobs are fast-paced, you work independently, there is lots of travel, and you are in a position to help a company achieve the next stage of growth and to reach their product development goals.

After riding out the aftermath of the 1987 stock market crash, I moved to a private leveraged buy-out firm. These firms specialize in buying businesses with borrowed money, in hopes of growing the business and realizing a big profit. After spending countless hours analyzing beauty parlor salons and meat packing plants, I realized that I had gone too far away from my first love—biotech.

BECOMING A CONSULTANT

The actual transaction to consultant occurred without much thought or planning. I had quit my job as leveraged buy-out specialist because of a difficult pregnancy—that ride from our condo to the office was making me turn green on a daily basis. While talking to one of my many friends in the biotech industry, Cynthia Robbins-Roth, I learned that she had just landed a major consulting assignment in agricultural biotech, an area that I was familiar with from my investment banking experience. Over the past 3 years, we had spoken regularly about biotech financing issues, and I had written some articles on the topic for her newsletter, *BioVenture View*. This interaction made me feel comfortable that we could work together. We teamed up on an informal basis and completed the assignment in about 6 months. This first experience set the stage for our continued partnership, now in its ninth year.

Cindy and I are essentially "freelancers," that is, we are self-employed and we run a small business. This experience is very different from working for one of the large accounting firms such as Bain & Co. or Andersen Consulting. All of my comments here refer to owning your own consulting business.

In our experience, most people who set up an independent consulting business end up taking a job at one of their clients. Why? Because despite the rewards, there are many drawbacks to setting out on your own. Many people miss having an organization around them and the certainty of a regular paycheck. They dislike not knowing from month to month what projects they will be responsible for or whether there will be a project at all! Some people are very comfortable with using their technical expertise, but don't like having to sell that expertise—the marketing side of the equation is important or you won't have much work.

We have overcome many of these obstacles and you can too, if you know what to look for.

If you decide to strike out on your own, here are some questions that will hit you early on.

What Are You Selling?

My partner and I both began working careers in the academic lab and in what were then early-stage biotechnology companies. I moved into the financial community, providing services to the industry; Cindy moved from the lab into business development within a biotech firm.

In all, we built up roughly 15 years of experience and industry watching before trying to tell CEOs and venture capitalists how to run their businesses. Our biggest selling point is that we have grown up with the biotech industry and we have been through several financing cycles. We have watched companies evolve from 2-person startups to public companies with hundreds of employees, and we have worked with management teams as they grappled with the challenges of this growth.

This experience, and the strong personal contact network built up during that time, is useful to a broad range of clients: CEOs with big pharma experience but no real exposure to entrepreneurial settings or working directly with scientists; academic scientists with no experience dealing with the financial community or business partners; potential investors hoping for insight to aid them in choosing the best opportunity to finance; big pharma looking for the best technology and team with whom to partner.

At BioVenture Consultants, we also are selling our scientific expertise, but more importantly we are selling our broad perspective of the drug development process, that is, how to take good science and turn it into a commercial product. As time goes by, our lab skills have become more and more antiquated, but our understanding of the scientific process is still very applicable. The ability to provide critical evaluation of potential new projects, to objectively review an R&D plan, and to help a client manage a scientific collaboration is a very important part of the services BVC provides. We also have honed our writing skills, a dying art in this country, so we can turn out a good business plan quickly. Communication is key, whether with potential investors, potential key employees, or with the business press.

Other areas where we have seen consultants flourish is in the clinical trial and regulatory area (backgrounds in the FDA or in clinical trial management), in public relations or investor relations (sometimes the two fields overlap), and in arranging partnerships between biotech and big pharma (consultants here usually have business development expertise).

Do You Take on a Partner?

Give serious thought to teaming up with someone. One of the biggest drawbacks to the consulting life is that it can get fairly lonely. If you are having trouble with a client, who do you call? There is nothing better than a partner who can share your woes (and triumphs!). In addition to listening to you, a partner can cover for you when you go on vacation, and, more importantly, he or she can tell you if your ideas are sound. A day-long meeting is much easier to handle if there are 2 of you. Being "on" for 8 hours during a client meeting is tough work, and a partner is invaluable during these times. Additionally, a partner with complementary experience and education can extend the areas in which your firm can operate with confidence.

But having a partner is work. Problems can arise when splitting up work (and fees) and handling the day-to-day affairs of the business. We have seen partners put together lengthy formally documented partnership arrangements, only to see them fall apart in a few years.

We have taken the informal route—there is no written document spelling out how we will work together. Each assignment is divided based on who has time available and the skills required. This approach has worked well because we had the luxury of time to work out our relationship, and we have a lot of confidence in each other's abilities and character.

The consulting partnership started as a part-time affair and was that way for the first couple of years, while our children were infants and Cindy got her publishing business more established. By the time BioVenture Consultants became a full-time occupation for both of us, in early 1991, we had collaborated on about a dozen assignments and we had figured out how to work with each other. Since we both have strong personalities, and we take pride in our work, it was important to find a way to work together and take advantage of our different strengths to create something better than anything we could have accomplished on our own.

This "getting to know you" attitude has been carried over into our consulting business. With each new client, we write a contract that includes a small "getting to know you" piece of work, which we bid fairly inexpensively. This gives the client direct experience with our working style and our capabilities. If the client is happy with that, we follow that initial project with much larger projects. This approach has worked extremely well for us. We are confident that clients will hire us for additional projects. This allows us to lose a little up front and to feel fairly confident that we will make it up on later parts of the job.

On a typical day, we will talk on the phone about 3 times, e-mail once or twice, and fax one or two documents back and forth. America Online has a great feature that allows you to send formatted documents back and forth, and we use it extensively. We have tried at various times to keep each other's calendars, but in keeping with our informal arrangement, it hasn't really worked out. We discuss important dates and we have learned to be flexible if one of us forgets to let the other one know what is going on. One of our greatest challenges was to learn to coordinate our activities so we could leverage our time.

We have toyed with the idea of adding someone to our partnership, but the right person has not come along. If you are just out of school, one way of breaking into the business may be to find a group that wants to add a junior person.

How Much Do You Charge?

The best way to find out is to ask people who consult in the same field that you intend to enter. Don't assume that all consultants are your competition. Most of us specialize in aspects of the business that are different enough so that we gain more than we lose from staying in touch with each other. In fact, clients really appreciate it when you admit that a project falls outside your area of expertise and you recommend another consulting group.

In the biotech management field, daily consulting rates range from $500 per day (usually for more inexperienced consultants) to $5000 per day (for larger groups that put several consultants on the job). Sometimes there are "success fees" on top of the project fee, typically when you help the company raise money or find a corporate partner. In some cases, you can get part of your fee in stock or stock options that can provide a chunk of cash if the company is successful.

The income earned by consultants is extremely variable, depending on how much work you bring in and the rate you charge. A consultant with around 10 years' relevant experience and a strong reputation should be able to earn from $100,000 to $150,000 per year. Keep in mind that you most likely will not be working 8 hours a day, 7 days a week; your fees have to get you through the inevitable downtime.

It is a real temptation to underbid your work when you first start consulting, because it takes time to learn how long certain assignments will take you and you just can't believe that anyone would pay you that much to do something that seems so straightforward. Remember, you spend a lot of nonbillable time adding to your experience and information base by following the industry, taking care of your business (paying bills, invoicing clients, fixing the fax machine), attending technical and industry conferences, maintaining your network of contacts, and generating new business. Your clients must indirectly reimburse you for maintaining your business. Also, issues that seem intuitively obvious to you, thanks to your years of direct experience, are not obvious to many clients. You must value your contribution if you want the client to value it.

You can bill on a "time spent" basis, or give a fixed bid for a project, based on your estimate of the time required to complete it. You can be compensated with a combination of cash and stock in the client's company (source of a potential upside if you do a great job and the stock value sky-rockets!). We currently bid most jobs on a fixed-bid basis simply because the client is more comfortable knowing how big a financial commitment they are making. After all this time we have a good idea how long a particular project will take. If you bill on an hourly basis, make sure to keep close tabs on your time. You will spend more time working than you realize, and clients will sometimes challenge your claimed hours—be sure you can account for the time you spend on a project.

There is a great deal of satisfaction gained when the check arrives in the mail. In a regular job, you can feel that you are paid regardless of how much or how little you work. In consulting, there is a direct correlation between the hours you put in and the reward. It is very gratifying to feel so in charge of your destiny. The downside, of course, is that the check needs to arrive sometime. Consultants moan about times of "feast and famine" and it's true. If you don't have funds to fall back on (retainer clients or a working spouse), only careful planning will get you through the dry times.

Getting It in Writing versus a Handshake

Over the years, we have developed a client contract that we insist must be signed before we do any work. You would be surprised at how many eager clients blanche when they see the contract. Just seeing it in black and white makes them get serious. Over the years we have found that clients sign about 75% of the contracts that we submit without significant changes.

Early in our business, it was difficult to demand a contract from clients, but now, if a client resists, we know that they are not serious about working on a significant project with us. It can become tense at times in the negotiation stage, but better so before you have invested any time rather than later when you submit the bill. "No surprises" should be the rallying call for consultants when dealing with clients.

If you would like to see a typical contract, call us and we would be happy to share it with you. Key components include a description of the work to be done, a clear outline of the time frame and fees, boilerplate legalese to protect you from being sued by third parties for something related to the project (this can be a real liability issue when you work on projects related to financings or deals), and limiting your liability to a maximum of your fee (otherwise, someone could wipe you out). The contract will not protect you in the event of gross negligence, breaking confidentiality, or illegal behavior.

A contract is also useful in dealing later with the IRS. It is important to define in writing your independence from your client so you can reap all the tax benefits (and Keough savings) of being a small business.

What about Travel?

Even for those who live in areas heavily populated by potential clients, travel is an important part of the consultant's job. We travel at least once a month, usually across the country, to visit current clients, to pitch business to a potential client, to attend a scientific conference related to a project, or to give a presentation at an industry conference. We have clients in Canada and all over the United States, and we travel to Japan and Europe once a year for conferences and to visit companies.

Establishing yourself as an industry expert by giving strong presentations is a great way to keep in touch with your network and to market your services—which essentially consists of using your brain. A great presentation on an area related to your services gives people a "free look" at your capabilites.

If you are traveling for a client, they cover all expenses. Make sure you keep good records, both for the client and for the IRS! Don't abuse your clients by flying first class (unless they agree ahead of time) or by charging nonessentials to their account.

IT'S ALL ABOUT REPUTATION

The only real asset of any consultant is her or his reputation. Most of our clients do not find us through direct marketing efforts (we don't advertise) but through recommendations by their personal contacts in the industry. We are reasonably well-known in the industry, in part because we are frequent invited speakers and moderators at industry conferences, but more importantly because we have worked directly with many key players. It is absolutely crucial to maintain ethical behavior at all times, to do your best job for all clients—even the ones who drive you crazy, and to maintain strict confidentiality. You will be privy to many pieces of information that, if disclosed, could materially harm your client—and could break SEC laws and land you in jail for insider trading.

There will always be some folks who just don't like your style or your "bedside manner." Don't worry about that—as long as you are clear that your work is high quality. Stick with clients who are comfortable with your personal style, and learn as time goes by what behaviors you might want to change. The relationship between client and consultant is very close while the project is underway, and trust must flow in both directions.

While I took a circuitous route to becoming a consultant, none of my former jobs was a waste. My philosophy is, "If you don't try it, how will you ever know that it isn't for you?" Working to earn a living lasts decades, so there is plenty of time to try different careers—especially while you are in your twenties or thirties. In the end, all my jobs have helped me directly or indirectly to build skills that eventually were put to use in BioVenture Consultants. O.K., so you want to know how the meat packing plants were helpful? Several of our biotech clients have been interested in acquiring various businesses, and my experience with both leveraged buyouts and investment banking has allowed BioVenture to lead these discussions. The experience gained while conducting these financial analyses is relevant regardless of the specific business sector—numbers are numbers.

In closing, the best thing about biotech industry consulting is that we are essentially paid to learn cool new science from some of the smartest people around. We get to work with a new cast of characters every month, and we are constantly learning new skills. Our network grows every day and we work hard to keep that growth going. We have reasonable control over our work load, and can take time to be with our families and to have other interests. Most importantly, at least for Cindy and myself, we don't have a boss telling us how to do our job. We are our own bosses. And science remains at the core of our business.

chapter

12

REGULATORY AFFAIRS:

Keeping Product Development on Track

Elizabeth Moyer, Ph.D.

Vice President of Product Development, Kinetek Pharmaceuticals, Inc.

In the world of shrinking numbers of research grants and funding, and fewer academic positions, there is a drive to find less traditional careers in science. In the last few years, there has been a dramatic decrease in NIH grants awarded to scientists under the age of 36 years.

Michael Teitelbaum of the Sloane Foundation, speaking at a National Research Council public meeting on trends in early research careers of life scientists, asked in the open forum whether it would make a difference to entering graduate students if they knew that they only had a 10 to 20% chance of obtaining an academic research position. Clearly, to some it would, but at the same time, this partly reflects the intellectual environment in academic training.

Many academic faculties have a negative attitude toward nontraditional career paths. In fact, early in my career I was told by one department head at a major university that they would only take graduate students headed toward an academic profession.

Current science training also does not direct students toward nontraditional career paths. Generally, the "best" students are steered toward graduate school, then on to an academic career, with little discussion of the other options available. Nor is there a visible pathway for young scientists who would like to choose less traditional careers.

One such alternate pathway is that of regulatory affairs. People go into regulatory affairs for about the same list of reasons people go into science itself. Some just want a job, some have more altruistic reasons (making a difference, helping people), and some are drawn into it and find it rewarding for various personal, intellectual, and professional reasons.

For a scientist, even one without an advanced degree, this field represents a reasonable career alternative to the grant-restricted world of academic science. In addition to using the philosophy ingrained as part of scientific training, specific skills and interests gained during college, graduate school, and/or postdoctoral work can also be critical.

While regulatory affairs can affect the quality of the food you eat, the cosmetics you apply, or the medicines you take, in this chapter I will focus most specifically on regulatory affairs as it applies to the pharmaceutical industry.

WHAT IS REGULATORY AFFAIRS?

As a discipline, regulatory affairs covers a broad range of specific skills and occupations. Under the best of circumstances, it is composed of a group of people who act as a liaison between the potentially conflicting worlds of government, industry, and consumers to help make sure that marketed products are safe and effective when used as advertised. People who work in regulatory affairs negotiate the interaction between the regulators (the government), the regulated (industry), and the market (consumers) to get good products to the market and to keep them there while preventing bad products from being sold.

The range of products covered is enormous, including foods and agricultural products, veterinary products, surgical equipment and medical devices, *in vitro* and *in vivo* diagnostic tools and tests, and drugs (which range from small molecules to proteins). The range of issues addressed is huge, from manufacturing and analytical testing, preliminary safety and efficacy testing, clinical trials, and postmarketing follow-up, to advertising issues, with a healthy dose of data management, document preparation, project management, budgeting, issue negotiation, and conflict resolution thrown in along the way.

Over the years, a complicated system of checks and balances has developed to set in place a process to efficiently and effectively regulate the marketing of products. On the industry side, people in regulatory affairs work with research scientists, clinicians, manufacturing groups, and sales and marketing groups to make sure that the government has the information it needs to judge a product. On the government side, people in regulatory affairs work to interpret and implement laws that Congress establishes to help protect the public. To carry out the congressional mandate, the Food and Drug Administration (FDA) requires pharmaceutical companies to generate and provide all the information deemed necessary to evaluate a given drug, biologic, and/or device with respect to safety and efficacy. This information is used by the Agency to decide whether the product should be on the market—and if so, how it should be marketed and sold.

On the consumer side, people in regulatory affairs help keep the other two groups honest and they provide the stimulus for Congress to enact the laws that regulate how government and industry treat products.

REGULATORY AFFAIRS IN A
BIOPHARMACEUTICAL COMPANY

Typical functions of a regulatory affairs group in a pharmaceutical company include interacting with regulatory authorities, preparing documents for regulatory submission, developing regulatory strategies, and interacting with company staff. Each of these responsibilities can have major implications on the success or failure of the company.

Interacting with the government is a crucial role within the company that is frequently restricted to the regulatory group. Whenever a company contacts the government, both sides typically document all topics discussed and the issues raised and answered. Funneling all conversations through the regulatory group prevents different people from making commitments to the government that are not fully known by the rest of the company. Typical topics for discussion include the planning of meetings, the format and content of documents to be submitted, the intended dates of submission of documents, the design of preclinical and clinical studies, changes to preclinical or clinical protocols, chemistry and manufacturing issues, adverse events (expected and unexpected) that occur during preclinical or clinical studies, labeling issues, international issues (permits, inspections of foreign sites, foreign regulatory actions), and other topics specific to the studies or products under discussion.

Preparing documents for regulatory submission is probably one of the most widely known functions of an industry regulatory group. Classic pictures of regulatory affairs groups often show them surrounded by steep piles of paper that cover dozens of tables as they prepare to submit an NDA (New Drug Application, asking for permission to make a drug available as a product). In fact, this is a monumental task (pictures don't lie), but it typically is preceded by years of previous submissions and months of work.

Most of that pile of paper on the tables was probably already submitted to the government in one form or another—at least if the company was smart. Realistically, this NDA, which usually represents about a decade of work at the company, has been preceded by dozens of previous submissions, meetings, and discussions, all of which are reflected in this huge mound of paper.

The filing of an application for a waiver to initiate clinical trials is critical to initiation of the process—in the United States, this is termed an IND (Investigational New Drug) application. Testing of a new drug in humans is only legal in the United States if the IND has been filed and the testing has not been put on clinical hold by the FDA. Clinical hold means that the FDA has concerns—usually for safety reasons—about the study design or about the preclinical work done to date. The format, content, and submission of this document are all done under the aegis of the regulatory group.

Following the initial IND filing, numerous other filings to the IND take place over the following years. Some of these are part of an annual update, others are filed separately as specific issues arise. During clinical trials, any adverse events are reported to the FDA. If the adverse reactions are serious, unexpected, or more frequent than expected, the events must be filed with the FDA within 3 to10 days of the event, depending on whether the event was fatal or life-threatening. Typically, these reports cite the nature of the problem, with background information concerning the problem and/or the patient, and they contain a summary evaluation of what this problem is likely to mean in terms of the potential risk to other patients still being tested with the drug. Dealing with this problem requires quick action and coordination with the clinical group and any other groups involved (preclinical and/or manufacturing), and is essential to building and maintaining the trust the FDA has for the company and its products.

Interacting with the internal company staff is an essential and sometimes difficult part of the job. It is the responsibility of the regulatory group to prepare reports for submission to the government that contain all the information necessary under the various regulations and guidelines, and to make sure that the data contained in these reports are accurate and verified. They must also make sure that key people within research and development have the necessary training to perform the tasks relevant to the drug in development, and that this training is documented, is complete, and is up-to-date. Often, basic researchers do not take kindly to what appear to be meddling paper-pushers bugging them about how to do their jobs.

The regulatory group also participates in the strategic planning for each group to help make sure that they operate within government guidelines and are able to provide the necessary information required at each stage of product development. Since the information required at each stage varies according to the nature and scope of the preclinical and clinical studies to be done, it also varies according to the particular government that will review the data. Because the requirements change over time, this can be a case of shooting at a moving target. This means that the negotiations required within a company to extract the necessary information and cooperation from people not in tune with government regulations during the product development cycle can be at least as challenging as dealing with the various governments that are reviewing the documentation.

Developing tactics for efficient product development and approval that can play a key role in the overall strategic plan for a company is another facet of regulatory affairs. Will the company file in the United States, in Europe, in Japan, or in all three? Although there is a movement toward harmonization in terms of the information required in these three very different markets, this process is not complete. Attempting to meet the requirements of all three regions can increase the cost of drug development sharply. Some companies enter into partnerships for foreign markets precisely because they do not want to grapple with these isses.

Some of the issues that must be examined include the kind of data that are critical for making the claims that will allow the drug to be marketed profitably and the minimum amount of data that must be gathered to support a successful filing. We must also ask whether these minimum data will support rapid development of the drug for other uses if the first indication falls through in the clinic. Is it more important to get on the market for one indication, then follow with supplements for other indications later, or should a broader use be claimed from the beginning? Should one go after an Orphan Drug indication (a situation where there are so few patients that the FDA gives 7 years of market exclusivity and special tax breaks to the developer)? Clearly, the answers to these questions can have a major financial impact on the companies. The wrong decision can delay marketing of the drug for years, or limit the market—and the profitability—of a drug.

Because the regulatory group is on the critical path to marketing products, this group is key to a company's success. While a company may have the brightest and best research scientists in the world working for it, unless it can bring to market the fruits of that research, the income stream for the company will be severely limited and investors will become very unhappy.

JOBS IN REGULATORY AFFAIRS?

The jobs available in a regulatory group depend on the organizational scheme for the group. This varies over time within a given company (and within the government), but in general there are five different kinds of responsibilities within a regulatory group: chemistry, manufacturing and controls, preclinical, clinical, and compliance and documentation. In some cases, one or more of these groups is autonomous, but has its own liaison person(s) to the regulatory group. In other cases, regulatory has specialists in each area who report directly to regulatory, while communicating with the respective group in research and development. These jobs, which are half-way in between the research and development groups on the one hand and regulatory affairs on the other, are obvious transition zones where a scientist within a company is most likely to be able to switch from a classical science job to a position in regulatory.

The chemistry, manufacturing, and control groups are responsible for documenting how the product and its raw materials are made and tested, and how the manufacturing processes are controlled. In the case of companies that work primarily with small molecules, these are people with experience in synthetic chemistry and classic analytical chemistry techniques. In the case of biotechnology products, these may be molecular biologists and protein chemists.

Proving that the a company's manufacturing methods are not only controlled but are also reproducible is a major issue with the government, which requires that these processes be validated. Validation is a series of formal tests undertaken to prove that each important step in the process can be repeated with predictable results. Even if the process undergoes changes, it is important to show that the product that results is identical. This means that all of the equipment that must operate within very specific temperature and pressure ranges has been appropriately tested and recalibrated as needed, that the software that controls critical steps has been demonstrated to run reliably on the equipment it controls, and that changes to the software can only happen under specific and well-controlled circumstances.

The output from each person performing these tasks needs to be reported and documented to meet government requirements. When a product is to be produced commercially, the government inspects all of the documents to ensure that they do prove that the process and product are capable of reproducibly meeting the product specifications during the storage conditions specified in the marketing application. Some of these jobs fall within the regulatory world of compliance (Did people do what they said they had to do to make the product?), and some fall within the realm of documentation (Do we have all of the right reports containing the right information in the right format for the market in which we plan to sell our product?).

Preclinical development does the work needed to get the product into humans in the first place and it provides supporting data that show that the product can reasonably be expected to be safe even if a larger population is exposed to the product than in the definitive clinical trials. Typically, this group studies the pharmacology and pharmacokinetics (absorption, distribution, metabolism, and excretion) of the product in various animal species, and it provides toxicology data in various species. The amount and types of testing required for a drug is the subject of ongoing discussions between the United States, the European Union, and Japan and it is one area where harmonization of requirements is most likely to reduce the amount and kind of testing required at various stages of the drug development process.

For the preclinical group, regulatory input on the progress of the harmonization process and on the specific requirements for products in each therapeutic area is extremely important. This group also generates reports that must meet specific government requirements for content. These requirements must be taken into account when preclinical studies are designed, otherwise more studies will have to be conducted to provide the necessary information for the IND filing. Typically, the regulatory staff and the compliance group within regulatory work hand in hand at each stage of the process for key studies to make sure that the study and the final report meet government requirements.

The clinical group oversees the clinical trials for a product. Depending on the organization of the company and the regulatory group, this group may report to regulatory affairs, or regulatory affairs may report to them. This is another key group that is capable of making or breaking the entire company. More money is spent by this group than by any other group in the entire company, and it is their work that decides the fate of a product.

As with the preclinical group, the work of the clinical group is highly dependent on a series of regulations designed to protect the test subjects. The regulations that define this work are subject to harmonization between the different governments, but the reality is that most countries prefer to have testing done on their own particular population group before they will approve the local marketing of a new product. This can lead to a flurry of regulatory activities, due to the regulations that control shipping the product for testing across international boundaries, and shipping patient samples back to the company for analysis.

The government is keenly interested in each step of the clinical trial process, including the development of the protocols, enrollment of the clinical testing sites, development of the data reporting forms, verification of adherence by clinicians and nurses to the protocol, and data reporting, data entry, and adverse event reporting. Each step is subject to government inspection and thus each is intimately involved with regulatory affairs.

Once the data are locked into the data base and data verification is complete, the statisticians take over. This group grinds the data into various analyses, looking for not only the overall results, but results by sex, by race (genetic background can influence the pharmacokinetics and therefore the safety and efficacy of a drug), and by population subgroup. This analysis is again subject to review by the regulatory group, as is the final report generated from each clinical trial.

The compliance group is responsible for making sure that the preclinical and clinical studies were performed as specified by a protocol that was written to meet government requirements. Depending on the organization of the company involved, they also review the chemistry, manufacturing, and control groups to ensure that their protocols, data, and reports reflect written procedures to accurately report the results of the various studies performed.

Documentation is the most obvious of all of the regulatory responsibilities. It can be the most time-consuming, tedious, and exacting of those jobs as well. People involved in this work assemble format, file, and store for instant retrieval all of the documents needed by the government to support the various filings required for a product. While this part of regulatory can be dismissed by the uninitiated as rote and as not requiring much skill, it is also true that compliance puts together the document that ultimately decides the fate of the product. If the reports are lost or unreadable, if the pagination and indices are mixed up or inaccurate, a decade's work can be put aside by the government until the company provides a document that is complete, comprehensible, and able to be reviewed.

For each of these areas, depending on the size of the company, at least one person is responsible for each task. In some companies, there are dozens of people responsible for each of these areas. Other companies use contract organizations to provide most of the staff, and have only a few inside personnel to make sure that the functions are performed as needed.

HOW TO SUCCEED IN REGULATORY AFFAIRS

The specific degree and area of specialty you bring to a regulatory affairs career will determine certain aspects of your career path, although almost any area related to science can fit into this world. Animal science majors, biochemists, clinicians, chemists, ecologists, entomologists, protein chemists, molecular biologists, pharmacokineticists, plant pathologists, statisticians, veterinarians, and virologists are just some examples of the kinds of scientists who find careers in regulatory affairs.

Clearly, entomologists are more likely to find work in the Environmental Protection Agency (EPA) or with a pesticide company than in a company that makes heart valves, and protein chemists are more likely to find a job in a biotech company or with the FDA than in a company that manufactures parenteral nutrient solutions. And while an advanced degree is less important in regulatory affairs than it is in academia or in basic research in industry, veterinarians are more likely to be put in charge of running large field trials of a new antibiotic for cattle than is someone with a bachelor's degree, and senior management positions are more typically filled by people with advanced degrees. Exceptions to these rules abound, of course, because demonstrated ability to get the job done is extremely important in this field.

So how do you know whether this would be a good career choice for you? The first thing you need to know is which regulations govern your product area. In addition, most people find that good people skills are important, and organizational skills and the ability to store and track information are usually critical.

An ability to see the broad view without getting lost in the details is another important skill, particularly for people aiming for more senior level management positions. A thick skin can also be an asset, particularly when acting as the bearer of bad news to top management. ("Yes, the FDA really will require us to do another $20 million clinical trial. No, the FDA won't let us claim our product cures everything in our labeling, unless we prove it.")

Flexibility, the ability to change directions quickly as new information comes in, and the willingness to give up cherished plans and theories are all strong assets in this career. You may think that you know what patient group will respond best to a new drug, only to find out as the trials proceed that the drug doesn't work at all well with those patients—but it does work spectacularly well in another disease. You may find yourself having to deal with last-minute manufacturing changes, trying to decide what else is going to change as a result. The regulations can change, meaning that a submission that was almost ready to go out the door suddenly needs to be reorganized, or extra studies must be done. Or, the company submitting the drug comes up with new and inexplicable results (good or bad) with implications that have to be determined quickly: do we stop ongoing clinical trials? Do we change trial design?

People skills are very useful in this kind of job. Product development— whether of a device, a drug or a pesticide—requires the combined input of whole teams of specialties. As a regulatory affairs professional, it is your job to gather information from across groups, and to help make sure that the story you are trying to sell to the government (or to industry) is complete and consistent. In some cases, this may mean coaxing a report in the appropriate format out of an unwilling investigator, or negotiating product specifications that meet manufacturing realities, clinical necessities, and legal requirements. You can find yourself in a hostile situation when different groups within your company or agency are fighting turf wars, but you are responsible as liaison to the outside world for making all seem calm and untroubled. You could also be the one who helps to get new laws passed to protect consumers, acting for your consumer group in a position where your ability to get the job done depends primarily on your powers of persuasion.

The ability to hear accurately not only what people are saying, but also what they are not saying, can be crucial. In working for the government, you may find yourself in a position where, because of your experience with similar products, you expect that a product may have effects clearly unanticipated by the manufacturer. In this case, you must get the company to do the necessary evaluations—but you may not be able to tell them why, because it would give them confidential information about a potentially competing product in development. As the parallel professional in industry, these hints can be essential to getting your product to the market quickly, and you need to know when to push for more information, and when to quit while you are ahead.

Organizational skills are very important as well. The typical NDA can be 200 volumes of more than 250 pages each, reflecting data acquired over 10 or more years of research. Organizing that information as it comes in, sorting it, storing it, and assembling it is a monumental job, requiring whole teams of people working for months. On the government end, reading and absorbing the information, quickly finding key questions, and coordinating the overall response to the company in the short timetable dictated under the new User's Fees rules can be tough. But when you realize that most companies are working against management goals of "X number of NDA's filed per year," you can imagine that the Christmas rush has a whole new meaning to the receiving docks in Washington. Keeping track of the various documents you are supposed to review, of what company responded to which query (and which issues are still outstanding), and preparing thoughtful written reviews of each application on time puts demands on anyone's filing system.

More and more, a talent for strategic planning is becoming essential as regulatory professionals are asked to assist in overall product development planning. Taking into account existing and in-process regulations and guidelines can help determine the probable development times and risks for a new product. The ability to successfully predict the quickest and most likely pathways to the marketing of a product has been known to make or break a young startup company that has only limited financial resources available to stay alive long enough to get a product to market.

Being in regulatory affairs has its downsides, too, as with any other job. An additional useful job skill is to be able to deal with the sequelae of being the bearer of bad tidings. Because regulatory professionals are the interface with the government, for some people within a company, they are viewed almost as being part of the government. Regulations passed to prevent problems caused by other companies sometimes seem to become the fault of the regulatory affairs group. Similarly, reports of compliance problems spotted by the company's own regulatory group can be greeted with ingratitude. The regulatory group frequently shares the blame when the government refuses to permit the marketing of a new product.

Never forget that over the years, you develop a reputation within your area. If you have the reputation for straight shooting and fair evaluation based on good science and backed by the facts, you are more likely to get things done in your own organization as well as with the companies or government agencies with which you deal.

How Do You Get into Regulatory Affairs?

If you ask 100 people in this field, and you'll probably get at least 75 different answers. The answer depends a lot on whether you want to work for government, industry, or consumer groups. Most people have the common denominator that they had some form of science background. But lawyers, MBAs, and the odd mathematician have been known to sneak in as well. It is not uncommon for people to start as a scientist in government and switch over to industry, because the "on-the-job" training in working with regulators is often highly prized by industry. While government jobs tend to be fairly low-paying, the initial investment in time spent training can pay off if you do decide to move to an industrial job or to strike out on your own as a regulatory consultant. Jobs in the consumer industry are less visible, and there are fewer of them.

Most commonly, scientists move into regulatory affairs as part of a career evolution. As a scientist, you can run into glass ceilings—jobs where no further promotion is likely. In some cases, career progression is blocked by the lack of an advanced degree. In other cases, your scientific specialty is simply not likely to lead to a high level management opportunity. For example, few toxicologists ever become officers of a company unless they leave their own narrow specialty.

Other people find that they prefer dealing with the wider spectrum of information—they find molecular biology too specialized, too out of touch with everyday reality. And for some, finding a science-related job that gets them out of the lab is a strong attraction. It is not uncommon for biologists to develop allergies to the animals they work with, and they are looking for some way to use their knowledge away from the bench. Don't count on a 9 to 5 job, however, particularly if you work for a small and growing company.

One way to move into government is to get a postdoctoral fellowship or the equivalent, working at the FDA, the Centers for Disease Control (CDC), or the EPA. There are various entry-level jobs in the government testing labs that can migrate to the divisions of the government that review marketing applications or that perform inspections. Jobs in the FDA are listed on its web site and are also listed in *Science* or in other scientific journals. People in the military can transfer to the Public Health Service, which staffs various positions within the FDA and the CDC. Carl Peck, a physician in the military who was very interested in pharmacokinetics, went this route and eventually became the Director of the Center for Drug Evaluation and Research at the FDA. This also can be a popular route for people who have had their advance degrees paid for by the government, and who owe the government service time as payback. The National Institutes of Health (NIH) is also a breeding ground for government regulators-to-be, since various clinical trials are conducted by NIH and there is extensive contact between the FDA, CDC and NIH.

In industry, there are also many routes. One common route is to start out in the lab, gathering data and writing reports for submission to a regulatory agency, learning in the process about the regulations affecting your area, then moving over into regulatory when a job opens up. Another route is to find an entry-level position as one of the people responsible for assembling and tracking the mounds of paper that are required to support products, then gradually becoming responsible for more strategic decisions on larger projects.

In companies that have project management groups (usually larger companies), scientists sometimes move over to these groups first, learning the wider scope of product development, before finally ending up in regulatory. In a small company, it can be by a much less logical route—the company needs to have something done, you're bright and available, so you are elected to get it done!

In general, the potential for switching around between jobs and developing titles and responsibilities that reflect your skills are greater in smaller companies, but larger companies provide more jobs and opportunities for getting a toe in the door. Logically enough, even small companies prefer to hire someone with a wealth of previous experience.

You can also start by getting a job for a contract research organization. These companies provide contract services to the pharmaceutical industry, by doing various parts of the product development process such as preclinical testing, analytical work, manufacturing, statistical evaluation, and clinical trials. Many of these companies have entry-level positions available. To find these companies, look on the Internet or look at lists organized by the various regulatory societies. One such list is "Pharmaceutical Contract Support Organizations" issued by the Drug Information Association (DIA). Once you have some training as a scientist who is responsible for generating data that are reviewed by the government, you can move over into regulatory affairs as a veteran who has survived (successfully) regulatory inspections and dealing with the government.

There are many courses and training seminars avaliable where you can learn the regulations. There are courses and seminars offered by societies involved with regulatory affairs, such as the DIA, RAPS (Regulatory Affairs Professionals), and the Food and Drug Law Institute. Each of these groups has a web page that lists upcoming seminars. You can also join these societies and get on their mailing lists—you can even join some of them on-line.

It is also true, however, that neither course work and seminars nor academic training can provide real-world experience. The diversity of scientific specialties in regulatory affairs reflects the simple fact that this is a career that requires on-the-job training. The best experience is gained through dealing with information that is typically highly confidential to a company; this limits the value of generalized seminars.

It is also important to get to know the people and personalities on the other end of the phone line when dealing with tough issues. No regulations can—or should—be extensive enough to fit all cases. Most decisions are made on the basis of what worked successfully before under reasonably similar situations. The dream product that nicely and tamely fits itself into the published regulations is a rare bird Creativity in interpreting and applying the rules and guidelines is often required when new areas of research begin to give rise to new modalities of therapy, such as gene therapy.

Regulatory affairs professionals occupy jobs at almost all levels within government and industry, from data entry personnel to the commissioner of the FDA. In industry, they rarely become chief executive officer or president, but senior vice presidents or executive vice presidents of regulatory affairs can be found in many companies. As stated earlier, it is not uncommon for people to choose this career simply to avoid glass ceilings. But if your aim is to become a corporate mogul, amassing billions and billions in personal fortune, regulatory affairs is probably not the best career choice. Regulatory affairs can, however, become a satisfying and rewarding profession.

One nice aspect true of this profession is that once you have gained a reputation for knowledge and professionalism in your field, you can strike out on your own as a consultant. This is intriguing for entrepreneurial types who like to boldly forge out on their own, marketing themselves and their skills to companies that don't really want to set up permanently the staff needed to bring a product to market. Sometimes consultants are also hired to bring in experience for a type of product that is new to the company, or when the company is looking for help in strategic planning.

The government sometimes brings in consultants to help solve a specific science question. Consultants roam from project to project, and Nancy Chew, a regulatory consultant for many years, has described regulatory consulting as being done best by someone who is easily bored. The nature of the job changes from project to project. It is not, however, for someone who is just starting out in the industry, because having a track record of success is critical in your marketing.

In summary, a career in regulatory affairs can be stimulating and challenging, and it can make extensive use of your scientific training. It requires in-depth knowledge and a tactical view of the regulations and the product, diplomatic skills, and a willingness to dive into details without becoming overwhelmed by them. But most of all, it requires a clear mind that is able to tie together all the scientific disciplines needed to make a product.

13

PATENT AGENT:

Protecting the Intellectual Property of Science

Pamela Sherwood, Ph.D.
Patent Agent, Bozicevic and Reed LLP

WHAT IS A PATENT AGENT?

I represent biotech companies and universities before the U.S. Patent and Trademark Office, and I advise them on intellectual property issues, primarily patent issues. About half of my work is done for academic inventors, and the other half for small biopharmaceutical companies.

Invention disclosures are brought to me in the form of manuscripts, short descriptions, and sometimes very short descriptions. The description of the basic invention forms the basis for a patent application. The application precisely describes the invention, and it also broadens the description to include features not originally found in the work, but that may have value in protecting the inventor's concept.

Applications are filed with the patent office, where they are examined. During the course of examination, I argue for the merits of the invention, and negotiate claims with the examiner that meet the requirements for patentability while protecting as broad a scope as possible for the invention. When foreign counterparts are filed to a U.S. application, I manage the correspondence with legal representatives in those countries. I work with an attorney to write legal opinions on the validity and potential infringement of issued U.S. patents. Finally, I sometimes help clients organize their management of intellectual property at the company. In the biopharmaceutical industry, intellectual property strategy is a very important component of overall business strategy and it plays an important role in a company's ability to raise financing and to build corporate partnerships with larger firms.

About 85 to 95% of my time is spent writing applications and responses to Patent Office comments with an occasional time-out for opinion work or client meetings.

As a preface to responsibility, I should note how law firms work. Associates or patent agents are expected to "bill" a certain number of hours in a year, typically between 1700 and 2200 hours, depending on the field of law and the particular firm, working directly for a client. Billable time does not include such things as keeping up with the literature, searching for lost files, reading recent federal circuit decisions, writing down time, managing a secretary, and so on. The expectation is that a quarter to a half of the time spent in the office will not be billable.

Work generally originates from a partner's clients, or from your own clients. It is a big advantage to have your own clients, because you do not need to rely on someone else to fill your day. In addition, the salary setup is such that someone billing the required hours will be profitable for the firm. Therefore, a person who brings in his or her own work will be economically desirable to a law firm.

Having clients is a mixed blessing, however. First, it requires some time and devotion to get and keep clients—this is a service business. The field is competitive, and certainly other firms would be more than happy to take over if a client is dissatisfied in any way. Also, it means a much higher level of responsibility.

The level of my personal responsibility has changed tremendously since I started working in this field in 1992. In the beginning, a senior attorney oversaw all of my work, reviewing everything before it went out. Since I did not have a federal registration number, I could not sign my own work that was submitted to the patent office. The clients came from a partner's group of clients, and so what I did was very fragmented, with bits and pieces coming from different companies and referring to different technologies and scientific fields. While this phase was unsatisfying in many ways, it was also much less stressful, because the ultimate responsibility for the work did not rest with me.

In time, and particularly after I moved to a different law firm last year, I built up my own client list and I generated all of my own work. This means that is I am not dependent on partners to give me billable work. The downside is that those clients expect constant attention, and I am the provider of that attention. When my daughter was born this spring, I could only be away from the office for a few days, because clients were relying on me to take care of their work. Someone had to look after the mail and phone messages, write applications when new disclosures came in, and so on.

In some ways the responsibility is terrifying, because no one is perfect, and I worry constantly about making a mistake that will have an impact on the position of someone that I work for and on the success of their company. This sense of responsibility leads to a very compulsive and obsessive life-style, where one worries about dates that might be missed, features left out of a description, breadth that was not sufficiently argued for, and so on. In spite of having done a thousand things correctly, it is the mistakes that haunt me forever. I am very proud that companies have trusted me with a valuable part of their business, but I would sleep better if they had not.

A DAY IN THE LIFE

Imagine staring at the screen of a computer with a word processing application running, and perhaps a handful of scientific articles piled up next to it. That pretty much sums up a typical day. I try to get to the office between 7 and 8 A.M., and I work straight through until 5 P.M., when I go home to take care of my two children.

A few phone calls come in; these are usually questions about applications that I am working on. I scan my e-mail for messages from clients regarding on-going work. A little time each day is spent talking to co-workers, including my secretary. The vast majority of the time is spent with my nose to the grindstone and my fingers to the keyboard, pounding out applications, revisions, and responses to patent office examiners.

Most associates and agents in law firms have their own office, as social interaction is not conducive to prosecution (the term for putting together patent applications and responding to the examiners). I leave the door open, but no one has much time to hang out and talk. It is a very solitary and rather monastic existence. A typical application takes 50 hours of writing time from disclosure to issue, with perhaps one or two hours spent on meetings and telephone time. This means a low ratio of "people time" to "computer time." Maybe twice a month, I meet with clients face to face. In most cases, this is not necessary, and telephones, faxes, and emails are enough to keep communications going.

Travel is nominal, although litigation assistants may spend a fair amount of time on airplanes. I go out of town a couple times a year, usually an overnight visit to a client. Occasionally, there is reason to visit the U.S. Patent Office and even more rarely, the European Patent Office.

WHAT YOU NEED TO DO THE JOB

Clear and precise writing skills are the most important skill for patent agents. Inventions have to be clearly defined, and it is my job to use words that will do them justice. Since so much time is spent writing, it should be clear that anyone who doesn't really like putting words to paper will not be suited for this job.

It is helpful to be able to put together a coherent telephone message. This sounds strange, perhaps, but in the wonderful world of phone tag, leaving coherent messages can cut down drastically on the time required to accomplish tasks and gain closure. One client noted that she liked my voicemail messages because I always told her everything that she needed to know, which allowed her to respond without having to catch me live on the phone.

E-mail has really overtaken the telephone as a key mode of communication in this job, so perhaps writing *is* everything. It also is important to have some management skills. Life is much easier and more efficient if you have help from a secretary or a docketing clerk, a file clerk, and so on. It is not so much managing in the sense of having a group to pull together, but convincing the staff to help out on your projects. If they don't like you and don't care about your cases, this makes life much more difficult.

Interpersonal skills are less important in the beginning, when you need only to please the partner with whom you work. Later on, it is critical to have client management skills; people should like to work with you personally. Because intellectual property law is a very competitive field, competence is not enough. It is very important that you have a genuine concern for the welfare of clients, and that you are easy to work. Again, keep in mind that this is a true service business. In addition, those people skills will come in handy when you have to help a client understand that they really don't have a good intellectual property strategy in place, or that their "brilliant invention" is actually not a legal invention that merits patent protection. You need to have their trust, to structure responses to the examiners, and to design a strategy for going forward.

Analytical skills are also very important. Legal thinking is not an easy switch for many scientists, because it lacks the concrete qualities of the physical world. Analysis is very abstract, and there are no hard right or wrong answers. If you are uncomfortable with ambiguities, then law is not your profession.

Scientific knowledge is helpful, but it is more important to know how to gather information from a variety of resources to assimilate new advances. No one can know everything, and disclosures will come in from a wide variety of fields. You need to manage a wide range of information sources—books, internet sites, technical papers, and symposia—and you need to keep a network of personal contacts going to be able to handle whatever comes your way.

Patent law tends to divide into prosecution—writing applications and prosecuting before the patent office, and litigation—filing lawsuits and arguing in court. The two can be very different in terms of qualifications and personality. Introverts who like to spend a lot of time alone end up in prosecution. Extroverts tend to do litigation.

GETTING IN THE DOOR

No one gets a Ph.D. in science with the idea that they will someday become a patent attorney. It requires too many years of long hours at the bench unless you really love it. My career began with a freshman biology course my first semester of college. The section on molecular biology was the most interesting thing I had ever learned about, and I immediately scrapped my plans for a liberal arts degree followed by law school (yes, there is cosmic justice in it somewhere).

By the end of college I was determined to continue a career in research, but the options at the time were quite limited—the only major biotechnology companies in existence in 1981 were Genentech, Cetus, and Biogen. In my senior year, I started on graduate research at University of Pennsylvania, my undergraduate college. One of the postdocs had a friend at Genentech, so I sent a résumé to California, and eventually I got a job there as a technician.

Genentech was an exciting place to be in the early 80s, but in order to get ahead you had to have a doctorate. In addition, I still wanted to pursue basic research. Stanford was close, so I applied for a Ph.D. program. Eventually, I chose a laboratory that worked in development of the immune system, combining two of my favorite topics.

As the years in graduate school passed, it became clear that, while I loved science, I hated bench work. It became painful to start work in the morning and put on another pair of gloves, pick up another pipetmen or another rack of tubes, or start up another tissue culture hood. This was combined with a desire to "have a life", which meant living in a house, earning more than $12,000 per year, and having children before turning 40.

I was looking for work outside of the lab, but with no success. Patent law actually came to me. My advisor had been involved in a start-up, and knew the patent attorney, Bert Rowland. Bert was looking for some help, so I faxed over a résumé. At that time, my personal deadline for leaving the lab was almost up, and I was seriously considering alternative careers—anything that could pay the bills. The offer to work in a law office was timely, and I literally had nothing to lose by accepting it. My original plan was to get some experience and then try again for a company job.

Much to my surprise, I found that I had a talent for legal writing, and stayed with that law firm for 5 years. It took a while for me to be accepted by the lawyers. For the first year, no one talked to me, except for Bert and his secretary. The clerical staff knew that I was not one of them, and the lawyers knew that I wasn't one of them, so I stayed in my office and wrote patent applications. It was lonely, but I got a lot of training crammed into a short period of time.

Some measure of acceptance finally came, probably because of money. My billing rate rose to the level of an attorney, and the firm loosened up to accept the fact that I could perform good legal analysis.

It is possible to find a job as a patent agent, although it is highly competitive. It is unlikely that a specific job will be advertised in a place that you are likely to stumble across, so you will have to rely on making contact with people in the business. The Internet is a good source, because many intellectual property law firms have web pages. A phone call is probably a good idea, rather than relying on a résumé mass mailing.

A better strategy would be to look up the local intellectual property law association—there are several in most metropolitan areas. These groups have numerous dinners and meetings, and this would be a good way to introduce yourself to local practitioners. Local law schools are another source, because large law firms recruit there. If you are still at school, the office of technology transfer could be another contact, since they will be working with local intellectual property law firms.

There is a huge salary range in this field. It takes a fairly big investment in training to get someone up to speed in patent law. Consequently, jobs for someone with no training are hard to come by, and do not pay very well. Starting salaries might be $35,000 at the low end, up to $55,000 at the top end of the range. The situation improves with time, particularly if you can develop your own practice with a nice client base. An established patent agent might make $80,000 to $100,000 at a big firm, and 40% of the billings in an independent practice at a small firm. The percentage of billings will depend on billing rate and hours worked, but it could be up to $150,000 (this is pretty unusual, though). Working as in-house counsel tends to pay less, on the low end of what a law firm would offer.

The training phase is not that distinct—the only real milestone is passing the patent bar, and that is typically done within the first year. Most people pass the exam if they have studied carefully for it. More to the point, there is a very long training period of gradually reaching some competence in all of the problems that have to be resolved. Two years of working full-time is the minimum to become useful to the firm and client, and to be able to handle most of what comes up. Most people do not have the experience to take on their own clients without about 5 years of experience. There is a lot to know, and it is not easy work, nor does it always make sense. Legal analysis calls for judging a situation based on knowledge of statutory law, case law, science, and day-to-day experience with clients and with the patent office.

There is no real promotion ladder as an agent. You cannot move up to being a partner at a law firm, nor can you become an attorney without returning to law school. Moving to a company or a university as in-house counsel is common for patent agents, as is returning to school for a degree.

Once you have that JD degree, you can become an associate at a law firm, a related position. For patent prosecution, the work of an associate is typically the same job as being an agent, except that the starting pay scale is higher and one can become a partner in the firm. In-house counsel can perform similar work, by writing and prosecuting applications, or she can manage intellectual property—this includes overseeing work that is done by outside counsel, negotiating licenses, working with scientists to get out disclosures, and so on.

Some law firms hire patent agents and some do not. It is not all that common, and most firms hire agents in conjunction with a program that sends them to law school, as part of a bid to recruit scientists into patent law. In my first law firm, I was the only agent among 30 lawyers. In my present position, there are two agents in an office with 40 lawyers.

I like my work very much. Interaction with inventors is interesting and sometimes fun. The work is intellectually challenging. I have a great deal of control over how to structure my day, although the flow of work tends to be overwhelming. Most of all, I like legal analysis and pulling together science and law. Also, it pays well.

I don't like the essentially dead-end aspect of being a patent agent. There really is nowhere to go from here without giving up a hard-earned practice and a lot of money to go back to school (a daunting prospect for anyone who has already made it through graduate school) to obtain a degree that does not make me better at my job. Billing is a hard idea to get used to. It is still a foreign concept to account for where every minute is spent, knowing that someone will have to spend a fair amount of money for that time.

14

· · · · · · · · · ·

FROM THE LAB BENCH
TO THE CLINIC:

A Career in Clinical Research

· · · · · · · · · · · · · · · · · ·

Katie M. Smith, Ph.D.

Director, Clinical Affairs, Gen-Probe, Inc.

My urge to leave the lab has taken me all the way to director of the department of clinical affairs at Gen-Probe, Incorporated, a biotechnology company in San Diego, California. Gen-Probe develops and manufactures *in vitro* diagnostic products for detecting infectious diseases and cancer, based on using gene probes to identify the presence of target nucleic acid sequences.

THE WORLD OF CLINICAL AFFAIRS

My work involves the planning and execution of clinical trials to support the licensure by the Food and Drug Administration (FDA) of new products for sale. I manage a staff with scientific backgrounds at the Bachelor's, Masters, or Ph.D. level in one of the life sciences—chemistry, molecular biology, biology, biochemistry, and so on. I supervise and guide my staff in developing clinical protocols, and in identifying medical institutions and clinics that would be good collaborators with us to evaluate our new products. For many of our products, clinical trials are required by federal law to provide information to FDA before that agency allows marketing of the final product. Once a clinical trial is underway, we conduct monitoring visits to each site to ensure their compliance to the protocol and to ensure that all phases of the trial are going well.

Back in the office, we receive huge amounts of data from the clinical trial sites, we enter it into a computer, and we analyze it, hoping that our product performs adequately to support the medical use we've targeted. At this stage, my clinical staff and its efforts interface closely with the regulatory function in the company to prepare a detailed report for the FDA on the product and the results of the clinical trial. This report is submitted to the FDA for their review and their eventual approval or clearance, at which time the company can begin to market the product.

My job entails meeting with members of my staff to review their progress in each clinical trial project and to troubleshoot any issues or barriers to successful completion of the study. In addition, I meet often with individuals or groups from the research and development groups, regulatory affairs, and marketing to address issues that can impact product development. Our group cannot work in isolation, but must stay in constant contact with the other teams within the company that contribute to product development and design.

My job and its associated activities require a broad range of skills for me to be effective and successful. Key job attributes include good organizational skills, clear communication—both written and verbal—good analytical skills for trouble-shooting and data analysis, and interpersonal skills to deal effectively with a myriad of issues and personalities both within and outside of the company.

In dealing with external institutions to conduct our trials, we confront numerous unexpected issues and, frequently, challenging personalities. To conduct an external study requires working closely with a principal investigator who may be an M.D. or a Ph.D., or possibly the head of a medical department and a leader in a particular field of medicine. This person often is used to being at the top of the totem pole, and is not used to having to respond to outsiders.

In addition to the principal investigator, the study really progresses as a result of the coordinated efforts of his or her staff, which includes research nurses, medical technologists, and so on. All of these individuals need different levels of training in the clinical protocols, along with attention and follow-up from me and my staff if our study is to be completed successfully.

All of this coordination, which is taking place simultaneously at several clinical sites across the country, requires a lot of time on the road over the course of the clinical trial. An even bigger chunk of travel time is spent at scientific meetings focused on specific disease areas in which our products are used, or meeting with the FDA (based outside of Washington, D.C.) to discuss our clinical and regulatory strategy for a new product. All of this travel takes up from 25 to 35% of my work time. While the travel can be grueling, I have some freedom to pick and choose the type conferences I attend. I actually enjoy the chance to step away from the daily grind in the office to gather a new perspective, through exposure to new people and ideas.

MY PATH TO GEN-PROBE

So, how did I get here? Was this a concerted, well-planned step in my career or was it serendipity? The first significant experience toward my job today started with my education. I had a strong propensity for science as I entered college and I received encouragement from my father, a chemist, which led to my early decision to major in chemistry. After college, I entered a Ph.D. program in biochemistry at the University of Arizona School of Medicine. This institution offered me a good mix of opportunities to satisfy my curiosity about biology and my confirmed love of chemistry. I entered into the first biochemistry class in the history of this institution, which consisted of 15 students. The small class size meant that we each received good instruction and lots of mentoring.

I discovered that those often fuzzy chemical concepts from my undergraduate education now became very clear and I relished doing laboratory research. It was both fun and challenging to have my own research project, to figure out how I could add to the scientific knowledge in the area and perhaps even make a significant impact.

I chose a project requiring the isolation and characterization of a protein in the blood coagulation cascade, based on an earlier project I had done in the Red Cross Research Lab. As is true of most graduate programs, I gained an excellent scientific knowledge of biochemistry, and an analytical and systematic approach to the design and execution of my experiments. An ability to write well in order to publish my findings was expected. In addition, and just as important, this project required a large amount of tenacity and perseverance as 50% or more of my experiments initially did not yield useful results. It seemed easier to disprove my hypothesis than to prove it.

Early in my postgraduate career, I met and married a fellow graduate student, whose focus was analytical chemistry. While the addition of a son to our family was welcome, it was also challenging with both mom and dad in graduate school.

Here again, tenacity and commitment were key. I was more focused and earnest about making both career and family work, and it paid off. We both completed our graduate degrees and our son survived us and the process quite well.

I always knew that I was not destined for a teaching career but instead aimed for an R&D job in industry. After a couple of years of postdoctoral training, I followed my husband to the Chicago area where he had landed a job supervising a small clinical reference laboratory that performed a variety of laboratory tests for doctors' offices and small local hospitals. Once our personal lives were somewhat settled and our son was enrolled in a local school, I joined my husband at the reference laboratory as director of research. Although the working conditions and culture were not focused on research, I found my job stimulating, with a variety of technical and personal challenges.

My job was to develop new laboratory tests in response to requests from our customers. My scientific background and training allowed me to explore the literature for published methods, to develop and validate one of those or to create a new method, to document it in a written procedure to be performed by the medical technologists working in the lab, and to train the lab techs to carry out the tests in a reproducible way.

This last step is where my interpersonal skills were challenged. I discovered that while my wonderful new assays had the potential to generate new revenue for the lab, the existing lab staff was only moderately enthused, at best, about learning these new techniques. Although I'd tried to design methods based on technology and procedures common to other tests currently performed in the lab, some new methodologies were required. I was excited about learning new technology and using new lab equipment, and I was rapidly expanding my scientific experience because each new test was so very different from the last. I discovered to my surprise that the lab staff was not very receptive to learning these new techniques because that meant a change for them. This change required taking time to learn new methods, which added to their work load. The addition of one more test to the lab's menu meant more responsibility for each of them in an already hectic work schedule.

Although I'd been convinced in graduate school that teaching was not my forte, I had to develop clear, concise teaching techniques and I offered lots of reassurance and enthusiasm to encourage the staff to accept and learn these new methods. While I initially thought I was contributing significantly to the organization, the initial reaction of those directly impacted by my work—the lab techs—made me feel that I was detracting from that environment. This experience expanded my technical expertise and taught me vivid lessons in how critical it is to deal effectively with people to gain acceptance for new ideas.

After about a year in this organization, I moved on to a very large pharmaceutical company, Abbott Laboratories, Inc., where I worked as a product development scientist in the Diagnostics Division. This group included scientists with backgrounds in chemistry, biochemistry, and immunology, and a group of engineers who were developing an instrument and chemical reagents for tests that would measure various cardiac, epileptic, and antibiotic drugs in patient serum.

My job was to to develop reference methods for these new tests using high performance liquid chromatography (HPLC). As time went by, I began to develop some new tests based on immunoassay technology. Again, I was able to apply my biochemistry and chemistry technical backgrounds to the job. I learned a great deal about organic and analytical chemistry as they are applied to real-world situations.

I also came to realize that I was fortunate to be in the right place at the right time. Everything we worked on came to fruition, either as a marketed product or by adding to our scientific knowledge about the technology and products with which we worked. I was also exposed to marketing and I participated in some road trips to visit customers and introduce them to our products and to do trouble-shooting in their labs. This gave me a close look at the customer. This early exposure was invaluable later when I had a role in recommending to company management new tests and new test features for development.

I also had my first exposure to clinical research. Our development team did not have a clinical department to do the studies and trials required for FDA clearance of our products. Consequently, we did a lot of this work ourselves. I had to identify test sites, train the site staff regarding our products and protocols, and analyze resulting data. I entered the world of computers and statistics, which was both necessary and extremely useful. This activity was very rewarding at the time—it gave me the opportunity to see how the real user handled the product I developed and how that product performed in the real world.

One of the other skills I came to appreciate and develop was the ability to multitask—to handle multiple, unrelated projects without dropping any balls. In this case, the tasks were to run an analytical lab and to develop immunoassays. These are technically very different functions that require different approaches. Also, the first task was a service function that I provided to several other scientists in the development group. This required that I meet their needs on demand while continuing to further my own development projects. As a consequence, I had to learn how to manage stress adequately. Both of these abilities have remained important as my career has progressed.

I was fortunate to have a scientific mentor in the form of my first boss, who still remains a good friend today. He ensured that I had adequate freedom to do my technical work as I saw fit. And even when I was challenged and sought his advice to solve technical problems, he helped me most by not giving me his best answer but by encouraging me to continue my independent pursuit of a solution. My mentor worked closely with the general manager of our project, who later left Abbott and moved to Allied Instrumentation Laboratory in the Boston area.

After 4 years at Abbott I was recruited by Allied as a project manager, overseeing the development of a small instrument and the associated clinical chemistry tests that were to be sold to doctors' office laboratories. This organization, like Abbott, included both chemists and engineers. I had to deal with both instrument and reagent technical development by guiding and motivating a diverse team of individuals.

A merger between our company and Allied Chemical, Inc., brought many months of uncertainty about our long-term fate. We experienced a lot of doubt and fear as to whether we'd have jobs the following year. Our concerns were well-founded—nearly everyone was faced with the need to find another job. Although I was offered a job right away at one of the subsidiary companies, I chose not to stay on the East Coast but instead, to find a job out West where I grew up and where my family lived. The disintegration of the organization in Boston accompanied the disintegration of my marriage, and I now had the freedom to seek the job and locale that suited me.

INTO THE CLINIC

So where does the clinical research experience come in? My next job moved me from research and development into clinical research. At this point, there were many people who, like myself, had worked at Abbott and later moved on to take jobs at other companies. I tapped into this network and found a path to the West Coast. I took a job at Hybritech, Inc.

My first job was as a research manager, not unlike the previous two jobs at Abbott and Allied. My job was to manage a group of scientists and technologists developing new immunoassay products for cardiac and cancer diagnostics. This company was fairly young, so a significant amount of discovery research was required for the company to understand its own technology and to produce reliable products. I was faced with the challenge of trying to further the technical development of each of three new products while learning the technology and quickly understanding the technical and organizational issues ahead.

I was brought into the organization at a middle management level. I quickly learned that others in the company felt that they should have been candidates for my job since they had been with the company for a long time and they felt that they had the qualifications. This did not make my job easy when interacting with these individuals. This was a new kind of interpersonal challenge that I hadn't encountered before. Dealing with and diffusing this resentment, in addition to the typical transition into any new job, required me to be in top form, both technically and managerially.

About 1 year into this new job and company, I came to seriously reflect on my job and career. Although my love for science and the product development application of science had not waned, I was sensing a need for a change. I started to explore other job functions in the company, which would allow me to apply my technical education and my prior experience, which would be fulfilling, and would not cause me to question if I were "selling out" on my academic training to be a laboratory scientist.

An opening came in the clinical and regulatory department, and I approached the departmental vice president, asking to be considered for the position. I got the job thanks to my past, although limited, experience in conducting clinical studies at Abbott and, more importantly, thanks to my analytical training at the Ph.D. level. Hybritech management held the view that a Ph.D. scientist working in clinical research was key to addressing the complexities of clinical trial design and data analysis, and could interact effectively with clinical investigators, most of whom were M.D.'s or Ph.D.'s. My technical training was expected to improve my interactions with these clinical collaborators.

Although Hybritech had been in business for about 10 years, the Clinical Affairs area was very undeveloped. I was required to learn all facets of the job on my own, and to build the department from the ground up. This included hiring new people and developing processes so that the new department could conduct good clinical trials. I knew that I could apply my skills in devising experiments in the lab to developing clinical trial protocols. However, I lacked depth of knowledge about the various disease areas to which our products would be applied and how to develop a specific clinical protocol for each product. A rapid learning curve ensued on my part. This came from reading the clinical rather than the scientific literature and in actually meeting with and talking to clinical researchers. I also quickly came to appreciate that a knowledge of FDA regulations and, more importantly, their interpretation, was important to the success of my clinical trial endeavors.

An amalgamation of my past experiences and of this new knowledge launched a different direction for my career change to clinical research. I also had an interest in people management, which over the next couple of years led me to position myself for a director-level position for both the clinical and regulatory departments. I continue to find both technical work and managerial work rewarding and challenging.

In my current position, I find that my basic education in biochemistry and an aptitude for systematic and analytical thinking have allowed me to handle most new technical areas or projects. I still do try to keep up with basic scientific literature and, of course, I continuously read the clinical literature in those medical areas that apply to our products.

I find it rewarding and renewing to interact with the research and development scientists so that I can stay current in the technical developments within our company and outside of it. This keeps me in touch with basic science. I've been exposed to some good role models in supervisors who have had strong managerial skills and I've tried to apply many of these skills to my own management style.

WHERE DO WE GO FROM HERE?

So what is the future for my career? I can see other opportunities to do the same type of clinical work but in another type of institution. Clinical research organizations (CROs) are for-profit entities that provide clinical trial services to pharmaceutical and diagnostic manufacturers. They do not manufacture pharmaceutical or diagnostic medical products.

CROs have staff members with clinical, regulatory, data management, and statistical experience who take on the clinical or regulatory work that a company does not perform on its own. This may involve hiring CRO personnel to monitor multiple clinical trial sites. The CRO may assist in the data collection process from each site and then they might perform the data analysis. Depending on the complexity of the trial and the in-house expertise of the client company, the CRO and its staff supplement one or more corporate functions to assist in completing the clinical development of a new product.

The jobs available at a CRO do not differ markedly from those within a company where clinical research takes place. Thus, similar positions and job opportunities are offered for individuals with science, computer, or statistical backgrounds. One key difference is that a manufacturer typically has a stream of products that are developed internally or are acquired from outside of the company, each of which requires clinical research before they can be brought to market. A CRO, on the other hand, conducts clinical research with whatever type of product a client brings to them.

As a CRO employee you may be asked to work on products in only one disease area in order to focus the CRO's expertise, or you may work on many products, no two of which are in the same disease area.

Compensation for a position with a manufacturer or with a CRO varies widely, depending on whether the position is within a pharmaceutical or a diagnostic manufacturing organization. Pharmaceutical companies typically pay more than do diagnostic companies for all levels of clinical trial personnel. CROs are competitive with pharmaceutical companies. East

Coast-based companies—manufacturers and CROs alike—typically pay higher salaries than do those on the West Coast. Salaries for entry level clinical research positions can range from $30,000 to $50,000, while those at the scientist or more senior levels range from $55,000 to $90,000 or more.

Good sources of training and networking to facilitate an entry or transition into clinical research work can be found in a number of societies. These organizations hold national and regional meetings once or twice a year and they offer courses and seminars on various clinical research and regulatory topics. Once such organization is SOCRA -Society of Clinical Research Associates. This is an excellent group within which to network since all of the attendees represent manufacturers or CROs that do clinical research work. Other groups of similar value include Barnett-Parexel, which offers many courses on the different facets of clinical research and these are scheduled on both the East and on the West Coasts, and the Drug Information Association (DIA), which also schedules national and regional meetings on pertinent clinical trial topics.

What It Takes

In looking back and also to the future, I've found that my analytical and scientific training and my ability to do critical thinking are of utmost importance in performing my job successfully. I believe that these skills have applied to all of the technical and managerial jobs I've held. In addition, communication skills, both speaking and listening, are important in making yourself understood and in understanding the input of others. Without good communication, things are not done efficiently and many interpersonal misunderstandings develop, creating tension and a negative work climate.

The ability to juggle many projects is key in today's working world, and this will continue to be the case in order to be successful in industry. Resilience and perseverance are handy traits that will help you withstand disappointment, confrontations, and criticism of your own shortcomings at times.

Starting with a good foundation in some type of scientific or technical training can position you well for a variety of jobs that are rewarding and challenging, and that will validate your decision to pursue an alternative scientifically based career.

chapter

15

· · · · · · · · · ·

TECHNOLOGY TRANSFER:

Enabling the Commercialization of Science

· · · · · · · · · · · · · · · · · ·

Susan L. Stoddard, Ph.D.

Offices of Technology Transfer and Strategic Alliances, Mayo Foundation

WHAT IS TECHNOLOGY TRANSFER?

Technology transfer is the process whereby inventions, discoveries, and other technologies developed at academic and research institutions such as Mayo Foundation are made available to the public by working with industry. Specifically, a technology transfer officer promotes the commercialization of an institution's intellectual property by transferring it into the hands of another organization—typically a company—that can do the product and commercial development. As the federal government funding of research has become more and more difficult to obtain, more academic institutions are looking to their technology transfer groups to provide alternative access to research dollars and income.

It is important to keep in mind that technology transfer is a process and not a single event. This process consists of many steps, starting with the disclosure of the invention or technology to the office of technology transfer. You can't develop a technology when you don't know that it exists. The step of disclosure is the one that is wholly dependent upon the investigator, although the technology transfer officer can facilitate the disclosure process by asking the right questions and by having strong interactions with the scientific staff.

Once the technology is disclosed to the office, the office must go through the processes of evaluating the technology and deciding whether or not it represents a patenable invention, making contacts with industry to evaluate company interest in the technology, establishing conditions of confidential disclosure with the companies that want to learn more about the technology, and with luck, finally negotiating a contract or license. This deal can range from setting up a collaborative research agreement between the academic scientist and a company, all the way through to a royalty-bearing license for a company to acquire all rights to an invention that it hopes to convert into a product. This process can take anywhere from several months to a couple of years.

The office of technology transfer at Mayo currently consists of six senior licensing professionals, each of whom has the authority to bind the institution in a contract negotiation. The aggregate professional degrees within this group are three Ph.D.'s, one J.D., two M.B.A.'s, and one Masters. Our group is supported by two licensing assistants, who assist the senior personnel in all aspects of the technology transfer process.

Mayo's office of technology transfer is a bit different from offices found in most universities, because it is run more as a business than as a service department. Our focus is on the financial bottom line, and we evaluate technologies based on their likelihood to return revenue to Mayo Foundation, not just based on their scientific excellence.

One policy that is notably different between our office and the standard university technology transfer office is that we do not patent every technology that is disclosed to us. Rather, we evaluate whether a patent is going to increase significantly the commercial value of the technology and we attempt to patent only in those cases. Part of the reason for this discrimination is that the cost to pursue patent protection can be very high, especially if you include the cost of foreign patent filings.

Determining the commercial value of a patent, which will be issued several years in the future with an unknown set of claims, is a bit like looking into a crystal ball. However, our office has built a successful reputation, bringing in slightly less than $12 million in gross revenue in fiscal year 1996.

HOW DID I GET THERE?

The training phase in technology transfer is likely to differ, depending on where you start your training. Because each of the technology transfer officers in our office at Mayo is really a senior position, there were no assistant steps that I could work my way through to move from junior to senior.

Rather, in my case, I was plopped down in the midst of an operating office and I was expected to begin making my contribution right away. In the first 6 months, I took a week-long review course for the patent bar exam to learn the basics of patent law and a 3-day licensing course through the American Management Association to give me an overview of the licensing process.

In truth, most of the learning process is an apprenticeship, in which you work with people who have experience in this area, learning how to ask the right questions and make the right decisions. Other groups that offer specific training in the area of technology transfer are the Association of University Technology Managers (AUTM) and the Licensing Executives Society (LES). In addition, local colleges offer a variety of business courses. You may also want to view "Techno-L," an on-line news group of university and industry people involved in technology transfer. This group has ongoing discussions about technology transfer topics, and it is a source of advice from colleagues around the country and throughout the world. It also provides information regarding changes in PTO (Patent and Trademark Office) policy and job openings.

Depending on your degree of worldliness when you first enter the business world, it is probably realistic to expect to spend at least 2 years developing your skills before you are comfortable in your new job, and another 3 to 5 years before you start completing multimillion dollar deals.

The promotion ladder, like the training process, will differ depending on your location. Some technology transfer offices may have a sequence of positions from assistant to associate to senior licensing person. The director of the section heads up the group of licensing professionals, but usually there is only one director per office.

The most common place to find a technology transfer office is within the university. These activities multiplied exponentially after the passage in 1980 of the Bayh-Dole Act, which contains provisions for the return of royalty income to any scientist whose federally funded research is commercialized. Prior to this Act, inventions and discoveries from federally funded research were required to be commercialized through the federal government—a process that was less than optimally effective.

With the passage of the Bayh-Dole Act, the field of technology transfer was born, as commercialization responsibility was shifted to the individual universities. Most institutions have technology transfer offices that were formed in the mid to late 1980s, and this continues to be an expanding field. An individual beginning in a technology transfer office at a university can anticipate a starting salary of between $40,000 and $60,000.

Industry also has technology transfer offices that reside within the business development department, in-license technology from either universities or other companies and out-license the technology that is being developed by the company itself. The technology transfer positions in industry tend to have a higher salary scale than those found within the university. In many cases, companies require that you have industry experience.

THE TYPICAL DAY

One of the seductive qualities about working in technology transfer is that there really is no such thing as a "typical" day; every day brings something new. In the university or in an academic laboratory, you have nearly total control of your time: planning experiments, attending or giving lectures, taking or administering examinations. In technology transfer you never know who is going to call on the telephone, send an e-mail, show up in a meeting, or otherwise require your professional scientific, problem-solving, or interpersonal skills.

Because each of the project managers works on all aspects of their projects, a typical day in our office consists of very diverse activities. There are always telephone calls and e-mails that need a reply or a follow-up; there are always letters or memos to dictate, edit, or sign; there is always the project or two at the bottom of the pile that is screaming out for you to find time to access the Internet or other resources to find companies to whom you might market this technology; and there is always a self-renewing stack of professional scientific, legal, or business reading.

Some days, however, are not ordinary. One such example occurred when I returned from a 3-day conference. The high-intensity meeting involved extensive networking and interactions with potential commercial partners. I attended several exciting presentations and I returned home with a stack of business cards and a list of contacts that needed follow-up. During the meeting I kept up with my audix messages and e-mail, and all was comparatively quiet and unexciting at the office.

On the night that I returned home, I checked my e-mail to find that disaster had struck! When I arrived at the office the next morning, I found my mail box stuffed to the top and the situation presented by last night's e-mail yet to be resolved. The e-mail had informed me of a business meeting scheduled for the next week, when it would be impossible for me to attend. This particular meeting had come out of nowhere, which is frequently the rule rather than the exception.

In this case, the investigator was discussing a technology that was partly imported from another university. We had not completed negotiations with the second university, and therefore had not fully secured the rights to commercialize this technology. Furthermore, the technology had concomitant commercial third-party rights. This was a situation in which the investigator did not fully understand the business details of the situation and planned to sit down in a meeting with a company where there would be no representation from our office of technology transfer. These matters needed to be addressed quickly.

On top of this, I received sequential phone calls from yet another university, and from our legal department. These calls related to difficulties in negotiating a material transfer agreement with the second university. We had finally decided upon the Universal Biological Material Transfer Agreement, to which both institutions were signatory. This document had been forwarded to the inventor, but since the agreement was unfamiliar to him, he refused to sign it. How were we going to deal with this one?

Just to make the day even more exciting, on top of the stack of mail sitting on my desk I found a memo from an institutional committee stating that they were considering returning the rights for an invention to an investigator. The problem was that our office had not supplied the committee with the appropriate information that would allow them to fully evaluate the situation. A reply to this committee needed to be formulated and drafted immediately.

The good news about "hair on fire" days such as this is that the they comprise only about 15% of all days. The bad news is that the 15% almost always seems to come in bunches, with several such days occurring sequentially. During this time, the mail box fills up, the e-mail screens overflow, and the audix system tells you that "your system is more than 70% full."

MOVING FROM LABCOATS TO BUSINESS SUITS

Everyone has a different reason for leaving the university. In my case, having been in the university all my life and risen to the tenured position of a full professorship, I was primarily bored. The way I found my present position was quite idiosyncratic, but I believe that is often the case when moving from science to business. My husband is on the staff at Mayo, so when I decided to look for a new career, Mayo was at the top of my list. My husband and I had had a commuting relationship for 7 years, and that aspect of my personal life was getting pretty old! Also, I was familiar with the institution, having collaborated with several individuals while on sabbatical leave and as a visiting scientist.

Now, however, I was approaching a scientific institution looking not for a technical position, but rather for one in administration. How should I present myself? Using my research training, I headed first for the library—this time, to find books on business résumé writing. I learned that the business résumé fits on one page. I was flabbergasted. How could I possibly condense my professional curriculum vitae, whose value is directly proportional to its length, into one page? This was a tremendous challenge and a very difficult exercise, extrapolating the real-life skill that I had developed over many years in academia "abstracting and evaluating" scientific literature.

Finally, with my one-page résumé in hand, I went to talk with all the administrators I knew (and some I did not) at Mayo. Without exception, they all pointed me toward Mayo Medical Ventures, which includes the office of technology transfer. At that time, by chance, the OTT was considering expanding its staff and hiring its first card-carrying professional scientist, who they thought might have some advantages over business people when interacting with the Mayo staff. My familiarity with the Mayo culture was an advantage, and I was hired.

Job openings in technology transfer are advertised through AUTM and LES, and they frequently show up on Techno-L on the Internet. Additionally, executive placement agencies may know of tech transfer opportunties in industry. One good source of job-hunting information is the chapter on "Recruiters Specializing in Biotechnology" in *1997 Genetic Engineering News Guides to Biotechnology Companies* (Mary Ann Liebert, Inc. Publishers; ISBN: 0-913113-79-4).

One aspect of the transition to business that I found difficult was moving from a position in which I was completely in charge and basically "knew everything," to a position in which I was an integral part of a working group, with people both junior and senior to me, and a position in which I basically had to learn from the ground up. It can be very frightening and frustrating to go from a field in which you have worked very hard to learn the material and feel that you are in control, to a position in which you are basically at sea. In the university, if you don't know the answer to a question, the one thing you do know is how to go out and find that answer.

I very clearly remember starting my new job in technology transfer and sitting in with my boss during a contract negotiation. I was astounded! How do you learn how to negotiate a business deal? So, I went to look for a book on negotiation strategy. Yes, there are books on negotiation out there; you will find when you read them that they are basically common sense, but notably short on specific "how-to" advice. This was a tremendous shock to me—not to have a source of black and white answers to questions. Rather, I found that much of business and negotiating is common sense, and the only way that it can be learned is through practice and being exposed to the various situations that can arise.

The transition from the lab to the business office can be challenging, exciting, frustrating, and sometimes just plain irritating. In the lab, you are generally your own boss—you wear what you want; you carry out your experiments when you want; you frequently speak as you wish with minimal on-line censorship; and you usually have some basic understanding of the politics and the pecking order.

In the business world, you wear "dress up clothes" every day; you are a bit more formal in your interactions with your co-workers and associates; you are expected to be at work regular hours, usually 8 A.M. to 5 P.M.; and you are set down in the middle of a notably different hierarchy. The hierarchy of people, positions, and personalities can be intimidating. In the university, most interactions are direct between the effector and the affected. One's department chair or mentor rarely intervenes.

In the business of technology transfer, other layers of interactions may become interwoven between the technology transfer officer and the inventor, especially if the inventor is unhappy. Some institutional and many office policies are best addressed by my boss, whereas other institutional policies are best addressed by my boss's boss. And with both very happy or unhappy inventors, there is the need to pass the message upstairs, so that the senior administrator is not blind-sided by a situation that should not have been news! Perhaps the business hierarchy is more accurately described as a web—most folks, like small insects, must be aware of all the threads they activate.

THE BIGGEST CHANGES

When asked what has been the biggest change for me in moving into technology transfer from the research laboratory and lecture podium, I have always answered that I have increased my short-term memory. That is not a joke!

In the laboratory you know what you are going to do when you arrive each morning; you move through the stages of an experiment in a sequential, logical fashion; and you usually do only one thing at a time. In technology transfer, you have to be able not only to juggle several tasks simultaneously but also to prioritize the existing tasks, sometimes with those priorities changing several times during the day. You may have a plan for the day when you come in, but the telephone rings, a Federal Express package arrives, the fax machine is smoking, your e-mail in-box is full, and you have three meetings scheduled back to back to back.

What was particularly difficult for me was to find a method to keep track of all the ongoing projects—basically a living, breathing to-do list that functions effectively and efficiently, and is something with which I can live. When I first started in this job, I would make a series of telephone calls, I would end up getting a series of answering machines, I would leave messages, and then I would be completely befuddled when the people called me back and I couldn't remember who they were. Hence, the need to improve that short-term memory!

Another significant difference between technology transfer and living in the university is the loss of control of the speed at which, and the direction in which, your projects move. Interacting with the commercial world can be very, very frustrating. A company that seems very interested in a technology and presses you to supply them with additional information may suddenly appear totally to lose interest, not calling for 6 months. Then, as the end-of-the-year budget dollars become available, they may be back on the phone to you, insisting that the deal be closed in the next 2 weeks or they won't be able to work with you.

Business trends, as exemplified by the daily changes in the stock market, fluctuate tremendously and your success at being able to commercialize your technologies must ride the crest of those fluctuations. The frustrations, however, are often balanced by the successes and the completed deals. I had a string of several weeks when every Friday, some time after 3 P.M., when you would think everyone was getting ready to go home and make plans for the weekend, the telephone rang with good news from a company regarding moving a deal forward or completing a negotiation. Go figure.

SKILLS FOR SUCCESS

Technology transfer depends on good communication: with the investigator to understand the technology that is being disclosed; with the patent attorney to efficiently and effectively pursue the prosecution of the patent; with the industry representative to effectively present the merits of your technology; and with the company's licensing professional or attorney to negotiate a mutually acceptable agreement.

Technology transfer is really a sales job in a fancy dress—you are "selling" your services to the investigator, who needs to be a satisfied customer, and you are selling the technology to the external company, who needs to be happy with the product they buy. Part of the interaction with the investigator may also involve informing him or her why that individual's pet technology is not going to be licensed or patented. This can sometimes involve fancy footwork and very effective communication, since scientists frequently confuse good science with commercializable science. Because they are in the ivory tower, scientists usually don't understand the intricacies of the marketplace and the bottom-line-oriented biases of the companies. The technology transfer officer is more likely to have a happy inventor if the expectations of that inventor are appropriately managed.

Because so much of the job of technology transfer involves interactions with individuals, you should be a bit of an extrovert if you are to enjoy your job. Developing relationships with investigators and companies involves varying degrees of small talk, as business discussions and negotiations proceed most smoothly when there is a personal relationship among the individuals involved.

You will also find that at professional meetings or trade meetings, business proceeds most effectively through a well-developed network. Such networks are constructed through friendly telephone calls, discussions over cocktails, and small talk at a meeting booth. An individual who is very introverted or who is uncomfortable meeting and talking with strangers probably would not find this a job that was made in heaven.

The opportunities to travel abound in technology transfer. However, due to the modern means of communication, including overnight mail services, e-mail, and telephone and video conferences, a great deal of business can be conducted without travel if you prefer that. I have found that the percentage of time I spend traveling has increased tremendously as I have become more experienced. This is in large part because with experience come bigger deals and negotiations, and as the importance of the negotiations escalates so does the value of travel and participating in face-to-face meetings.

Traveling can be very stressful and each individual needs to develop travel strategies that are effective for him or her. Laptop computers and audix systems allow you to keep in touch with the office while you are away and they make life considerably less hectic when you return. Since business travel also requires hotel stays and frequent business dinners, your common sense and good judgment are brought into play regarding what are appropriate business expenses. Clearly you need to maintain professional standards and behavior while still entertaining a client. Such decisions come more easily to some people than to others and they can be very stressful if this is something with which you are not comfortable.

THE RIGHT DECISION?

I have frequently been asked whether I have regretted the move out of the ivory tower into business. The answer has always been an unqualified "No!" Granted, I left the university at the peak of my career there, rather than at the beginning. Surely, the professional self-assurance that I had at that time aided me in the transition to an entirely new world.

Nevertheless, this new life has been challenging. But with the challenge has come very real excitement and reward, most particularly related to learning and expanding one's horizons. I have needed to become conversant with many aspects of business and patent law, financial analysis, and all sorts of science that was completely new to me. I was trained in and taught neuroscience, but since I work with inventors in all aspects of biotechnology, I have learned bits and pieces of entirely different fields, including immunology, bacteriology, parisitology, and molecular genetics.

It is a reward to work with other scientists who are experts in their fields and who are delighted to share that expertise. There are days when frustration and stress make me ask what I am doing in this fast-moving and dynamic job. But the satisfaction of closing a deal and moving an invention into the public sector always provides the answer.

16

· · · · · · · · ·

CORPORATE
COMMUNICATIONS:

Helping Companies Sell Their Story

· · · · · · · · · · · · · · · · ·

Tony Russo, Ph.D.
Co-founder of Noonan/Russo Communications

Often, scientists will tell me they are tired of being in the lab, but love the science. How can they work in the field, without having to work in a lab? How can I communicate the science without being the one conducting it? One great answer is to enter the public relations field, working with technology-based clients.

Five years ago it was unusual for recruiters in the public relations profession to receive resumes from scientists, much less from Ph.D.'s. After 8-plus years of advanced study, why would someone switch to a professions where the typical participant has only a college degree? I have spent the past decade as one of the few Ph.D.'s in public relations, feeling like a misfit in a profession dominated by people with undergraduate degrees in English and journalism.

Suddenly the climate has changed. I receive résumés from several PhD.'s each week. Today, people seem more inclined to investigate alternative careers and move into professions that don't appear to directly utilize their graduate degrees.

Alternative Careers in Science
161

For me, the realization that I could consider an alternative to further work in the field of psychology came when I was studying and conducting research at Harvard University. I saw my colleagues struggling to obtain even an entry-level job in a profession where they had spent 5 or more years studying at the advanced levels. These were bright people with a string of publications to their credit and short-term expectations no grander than landing a temporary adjunct professorship at a midtier state-run school.

My own hope of supporting myself with government research grants was cut short by the budding trend toward cost cutting at the federal and state levels. And so, my entry into public relations occurred not by design, but rather by the need for employment. In fact, when I received my Ph.D., I hardly knew the PR profession existed.

I found the route to a public relations career during a 2-year stint on Wall Street. I certainly wasn't the only Ph.D. (or degree holder of any level, for that matter) to be tempted by Wall Street in the early 1980s. We all thought that Wall Street might be an interesting alternative to research (to say nothing of the money you could make, significantly more than as a post-doc). None of us had much in the way of direct qualifications for being on the Street. The common wisdom was "if I can get a Ph.D., then surely I can do a spreadsheet." Besides, there were a few role models successfully practicing on Wall Street, many from among my group at Harvard. Why not practice along with them?

My personal role model was a practicing psychiatrist from Yale who headed up a major bullion trading company. Why did he leave psychiatry? As he was fond of telling me, "How much money can you make in an hour, $100? In private practice you are limited by the amount of time you can work. In business you don't have those limitations."

Although money was not my motivation, my role model's ability to switch professions, for whatever reason, made me realize that I, who had invested less academic time in my pursuits, might qualify for a career change consideration. Role models become important in switching professions because they give you a yardstick to measure yourself against. In academe or in the lab, you simply look at your colleague across the hall. In public relations, the only way that you find out about others' academic training is through asking. After a couple of years in what I now consider to be my second graduate degree, I found myself enjoying writing, research, and organization, three skills I needed to get my Ph.D. and three skills that are critical for success in public relations. The pay was poor at that point and not what I had fantasize—around $12,500 per year, not nearly enough to survive in New York City! But I liked the work and I was learning.

My first real introduction to PR was an undergraduate night course in public relations I took while I was on Wall Street. I applied what I learned in the course to my Wall Street job, and quickly realized I had found my calling. I could research companies, learn about a lot of different professions, write, and organize. PR turned out to be a profession that was stimulating, interesting, and that had job openings! For me, the ability to later work alongside of the health care profession, the very profession I had trained to become a member of, was an added incentive to make this move.

The professor of the PR course took me aside one evening and said I was wasting my time by working at a single private company and that to grow and develop in the field, I should work for a PR agency for at least 5 years. This would allow me to gain a variety of experiences working on several different accounts. She said I was talented, but that I really needed further exposure and she offered to point me in the right direction and write a letter of reference.

I took her advice and began to apply for jobs in the financial public relations field. Remember, I was coming from Wall Street and there were basically two specialties in PR: financial and general interest. At least I had a little experience to become a specialist!

I ended up at a small entrepreneurial financial PR firm that specialized in international companies. There we represented foreign companies doing business in the United States. It was a specialty that was being pioneered by the head of the agency.

Within 2 years, I saw this tiny agency grow to have offices in London, Brussels, Sydney, Hong Kong, and Tokyo. The rapid growth was a result of the leadership of the young CEO and the fact that the company had established a niche market. So, I learned lesson number one—put together a unique niche strategy that requires fairly high barriers to entry (in this case it was knowledge of world capital markets), specialize, then aggressively establish yourself in the field. With that important lesson learned, I moved on to gain more experience at other agencies.

I seemed to end up at agencies that were young, entrepreneurial, and in niche market areas. The agencies were all growing rapidly. Now there were more specialties such as real estate, health care, entertainment, and so on. I tried my hand at real estate, then at health care. By this time I had developed a real skill—I was good at organizing and I seemed to have a knack for understand the media—its needs, concerns, and politics. Also, I was good at putting a story together—seeing news when perhaps others did not, and more importantly, knowing when a story was not newsworthy or what elements it would need to become newsworthy.

But health care excited me. I was deeply interested in research and to have the ability to understand it, to synthesize it, and to place it in a broader context was a challenge. I had spent several years studying medical psychology at Columbia University and at Johns Hopkins and maybe now I could use some of that training in this new area of specialty. Maybe my life was coming full circle.

Then biotechnology was born. I had the good fortune to be working for an agency that had a biotech account (those were the days when there were only a handful of biotech companies). I began to pioneer a new area of PR: health care and biotech.

Success came quickly. I loved the work, I knew what I was talking about, and I quickly became one of the few experts in the field.

It was at that point that opportunity struck. My colleague across the hall, Susan Noonan, asked me if I might be interested in striking out—building a new firm with her. I realized that I had all the ingredients I needed to be successful: a skill, knowledge of medicine, and schooling in entrepreneurship from my previous positions. It was the beginning of the biotech movement and I was in the right place at the right time.

Noonan/Russo quickly established itself as a leading firm in this area. We were specialists and we circulated in the biotech and health care industries. And now nearly 10 years later, I have candidates for jobs, not unlike myself, sending me résumés.

So, what are the steps to break into the communications business, now that we Ph.D.'s don't have to be pioneers anymore?

Making the Move

Many of the Ph.D.'s who now come through my door at Noonan/Russo looking for jobs and advice are not as fortunate as the undergraduates I see who knew from their freshman years that they wanted to work in public relations and who have methodically sought out PR internships each summer. The Ph.D.'s are smart, but they are largely uncertain about what they would like to do with their life outside of the lab. As someone who has attained the highest academic degree possible, they usually have a strong sense of ego and are accustomed to working independently. This can make it tough to make the shift to working as part of a team, alongside of people with more direct experience but fewer academic accolades.

All of our clients are businesses related to the biomedical field. This means that our employees need to understand the technical implications of the products and services we are presenting to the press and investors, and they also need the communication skills. I often have to choose between an applicant with a strong PR background and someone who has a good technical background in the biosciences but lacks the essential PR skills needed to be successful in an agency setting.

In my business, one must know how to write a news release and how to write for a nonscientific audience, how to construct and write a public relations program, one must have a basic understanding of chemistry and biology, and one must feel comfortable in communicating with the Wall Street community. It is a large skill set and one that few applicants have when they interview for an entry-level position with an agency such as Noonan/Russo. Whether these skills are acquired on the job, in school, or through natural talents makes selecting the "right" candidates difficult.

So for me, there is always a gamble that the person I hire will be motivated to acquire the skills they don't have and use the skills they have to add value to our organization. A Ph.D. almost has to forget that they have an advanced degree, since they must learn a new set of skills to work in an environment where many may be threatened by their academic training. At Noonan/Russo, we divide our staff into several groups: those who are interested in the investment community (investor relations), the press (media relations), those who like to write (corporate communications), and those who are interested in marketing (product communications). Each group requires a different set of skills and a different way of viewing the world. For example, below is a sample of some of the questions members of our interviewing team might ask.

Media Relations

To join the Media Group we require a basic knowledge of the working media. This means that you must understand:

- What are the major publications important for client coverage, and why are they important?
- How do you write a news release, what goes into the lead paragraph, the first paragraph, the second paragraph, the boilerplate?
- What is an advance and an embargo? When do you move news? What are the laws regarding public disclosure? What is the difference between not for attribution, off the record, and on the record? What is news?
- Who are the important reporters for the different sectors, and what are they writing about? When do you call a reporter? What do you say? What do you say to a reporter who doesn't want to be bothered talking to you? What if the reporter hates talking to PR firms (learn not to flinch at the word "hack")?

Investor Relations

Two of the most critical issues for any company are how to raise financing and how to keep shareholders happy. A strong IR group can make a significant difference in a company's long-term success. Here, the importance of knowing how Wall Street operates is critical:

- What is the differences between a balance sheet and an income statement? What is a burn rate?
- Who are the major analysts that cover biotech, and which companies (Your client? Your client's competition?) do they write about?
- What are the major investment banks that finance in your client's sector?
- What are the leading companies in your client's sector? What are their products that might compete with your client's? How are they viewed compared with your client?
- What is the difference between the buy side and the sell side? What is the difference between retail and institutional investors?
- What are the trends driving the financial markets?
- What are the SEC regulations covering disclosure? What is the "quiet period"?

Product Communications

In the biomedical arena, product news is critical to a company's stock price, which in turn drives the company's ability to raise future financings and its ability to complete product research and development. Helping clients communicate clearly and in a timely fashion about their products in development is an important role. Some of the key issues include:

- What are the FDA rules regarding clinical trial result disclosures?
- What are the steps to FDA approval?
- What are the FDA advisory panels? How do they work?
- What is the difference between a PLA, a 510K, and an NDA?
- How do you launch a product? What are you allowed to say about the product and its competition? What isn't allowed? FDA regulations about disclosure can sometimes interfere with SEC disclosure requirements.

Corporate Communications

To keep shareholders on board, and to gain new shareholders (as well as to fulfill SEC requirements) companies must maintain an ongoing stream of communication with the outside world. Some of the important issues for a PR firm working with such a company are:

- What are key corporate messages? This can be amazingly difficult for a company to verbalize, but it is critical, especially in a sector where there are 250 other companies competing for the attention of the press, Wall Street, and investors.

- How are these corporate messages best expressed to different audiences?
- What are the best tools to use: multimedia, internet, print, and so on?
- What are the elements of good corporate design, and why is it important?
- What makes a good corporate identity?
- How do you judge the right style of design to use, and so on.

Naturally, we look not only at how well one can answer the above question, but also at one's portfolio of existing work. How have the job candidates demonstrated that they can perform the above tasks? What is the quality of their previous work? How can their references support their candidacy?

How Do I Break into the Profession If I Don't Have These Skills?

Get some PR experience: work as a volunteer, take PR courses, or try to get an internship at a firm. Imagine that you are still in graduate school. Don't focus on a salary, just get the experience so that you can demonstrate to a prospective employer that you want to be in public relations and that you have enough skills to be able to be productive from day one. It is perhaps the most difficult job you will ever try to get, since few organizations want to serve as a training ground for someone who has a little working knowledge of their profession—or even worse, for someone who is unsure of what they want to do with their life.

You've Got the Job, Now What Can You Expect?

The best employees we have had are always the ones who can't seem to learn enough, who are continually curious. PR, unlike other professions, is a bit unstructured. Anything can happen at any moment. The stock market can take a dive, a product can fail in clinical trials, a researcher can commit fraud. And it is the PR professional who will be called upon for advice and quick action. One needs to be organized, calm, efficient, and one must attend to details for multiple clients, usually simultaneously.

So what is a typical day like? It can begin with a news release, or several news releases, which can mean getting up in time to distribute the release as early as 3 A.M. Remember, some firms operate on European time. When it is 8:30 A.M. in Europe, it is 2:30 A.M. in New York. Sending a news release might involve faxing it to a group of publications you have chosen to get that news. This group may vary, depending on the news you are distributing. For example, a science discovery would go to a different group of reporters and interested individuals than would a personnel announcement. Besides going to a different list of publications, it might go to different reporters at the same publication.

Once you have sent out the news release, it is time to follow-up with the group of reporters to whom you have sent it. Are they interested in the story, and will they report it? Was there an interesting aspect of the announcement that they may have missed? Why was the paper so important? What role does the research play within the current body of research on the topic? In short, why is the paper important and deserving of coverage? Although many of these questions may have been answered in the news release, the role of the PR professional is to illuminate, to help educate the reporter about your client, to provide access to your client, to answer the reporter's questions, and to make recommendations as to whom the reporter might speak with at the company.

If you work in the investor relations area, you might call a Wall Street analyst or a mutual fund manager about the news, particularly if the announcement was made by a public company. If that person were convinced that the news was significant, the analyst might issue an internal note to her firm about the announcement and perhaps she might recommend purchase of the stock. She might also issue an external report supporting the same conclusion. If she is a fund manager, she might purchase the stock.

But what if the announcement is negative? What if a clinical trial has failed? To be prepared for negative news, or crisis management as it is called in the PR business, is a critical component of public relations. Develop a plan—before it happens, understand your different audiences, what their needs might be, and how to provide them with timely information. Your audience might be the company's employees, its shareholders, those who write about and follow the company, community officials, and so on. The message might be different for each audience. And while the temptation might be to put a positive spin on the announcement, one of the most important roles of the PR professional is to get the news out quickly and accurately.

In addition to the news that drives your day, a PR professional might write a brochure, devise a PR program or communications strategy, write a slide show for the investment community, or prepare a client for a meeting with the press or with the financial community. Your day could include speaker training, media training, or writing and producing a video. It could also involve finding a name for a new company, renaming an old one, or helping to create a corporate identity or logo.

In terms of a skill set, the public relations field can allow you to use your creative skills, your writing skills, your scientific knowledge, your knowledge of the biotech field, and so on. It is a field so varied, requiring so many different types of skills, that it is nearly impossible to be perfectly trained before entering the field.

SALARY AND JOB TITLES

Entry level positions start at around $25,000 for most large urban centers. A hard-working employee can expect to advance from this level within a year of starting the job. At the vice president level, the salary will be greater than $100,000.

At most agencies, progress is measured by client satisfaction, the ability to generate new business, and talent—producing creative work. Because PR involves a lot of juggling, you must maintain an air of calm at all times. The ability to multitask is important, and one must do this while paying attention to the many details inherent in the position. As you advance, you become responsible for larger accounts with bigger budgets and with increasingly sophisticated communications programs.

The alternative to working at a public relations agency is to work in-house at a pharmaceutical or biotech company. An in-house position may have many of the same responsibilities as does an external public relations agency. The job might involve responsibility for internal communications or product communications (news releases, communications with the media, etc.) with a greater emphasis on helping to position the story so that the PR agency can take it to a wider audience. The internal tasks might be divided into investor relations responsibilities and media or corporate communications responsibilities. Tasks such as writing might fall to the in-house person.

In smaller companies, you may need to be cross-functional. In larger organizations, an internal communications position might exist. Public affairs, crisis management, and governmental relations may also fall into the realm of the communications group. A large pharmaceutical company might employ 60 people in the communications area. The main difference, however, between an agency position and one in-house is that at an agency, one usually works with many clients on a number of projects. As a result, the in-house person often has a greater understanding of the company that she is working for than might an outside person.

In an agency, most newcomers begin as an account coordinator or junior account executive. If the candidate has experience in the communications business, it might be possible to begin as an account executive or senior account executive. In a biotech or other small corporation, one might begin as a manager of communications or as a writer.

Regardless of the position, you might take a pay cut if you are coming from academe. When I started in the business, I was told, "Your Ph.D. doesn't matter to us, you have to start at the same level as everyone else." So my first few years were viewed as further graduate training. It is important to focus on what you learn in the first few years, and not on the salary. If you are good, your scientific training will give you the push you need to move up the career ladder to the assistant vice-president and vice-president level.

TRAINING

Training in public relations is a daily activity. There are few formal courses that one can take. Until recently few schools even offered courses. Now one can get accreditation in public relations and can even be licensed by the Public Relations Society of America (PRSA), the trade group governing the field. To become licensed, one must work in the public relations field full time for 5 years and one must pass rigorous written and oral exams. Because the accreditation was introduced in 1964, and because the industry has historically not pushed accreditation, few professionals are certified. Although the PRSA has made great efforts to license people in the profession, this has little impact on one's ability to land a job or an account. However, if one wants to rise in the PRSA hierarchy, licensing is critical.

But what if you have not had any formal courses and you are fresh out of a doctoral program? How long will it take for you to gain a working knowledge of the field? The answer is about two years. In that time, you may write and distribute many news releases, annual reports, and brochures. You may produce slide presentations, set up analyst meetings, and perhaps organize media tours or investor road shows. You will work on Web sites, navigate through a crisis or two, and write many public relations programs. The result is a feeling of confidence that you will be able to manage most public relations tasks. As you advance through the hierarchy, you will be able to handle more and more responsibilities. The more experience you have, the better you will be able to advise a client.

MOBILITY

It has often been said that working for an agency is a critical professional move in a public relations career because it gives you a wide variety of experiences that will help you make more educated judgments. The skills and experiences gained from working in an agency will be useful in almost any PR setting. Working for clients on different programs in various situations gives you a wide range of life experiences (and great juggling skills). In-house, there is less variety and the focus may be on just one product. This is the difference between being a generalist and being a specialist.

As a specialist, you might work only with the media on biotech accounts. As a generalist, you might work on investor relations, media relations, or you might even write annual reports for a variety of firms that range from financial service companies to entertainment companies. It is like the difference between being a liberal arts major in college and choosing a major for an advanced degree. While larger PR agencies prefer generalists, the smaller, boutique agencies and the corporations prefer the specialist. To become a specialist will make you very employable in a single profession. However, if there are few jobs in that field, you might have fewer options.

As a Ph.D. you will find it easier to get a job as a specialist, because you have a specific area of expertise that few other people have in public relations. In fact, this is your key selling point. If you are a Ph.D., you probably enjoy the field you have specialized in and you want to stay close to the profession.

WHERE DO I GO TO FIND OUT ABOUT PR?

There are many directories where one can learn about public relations companies and their specialty. The most famous of these is *O'Dwyer's Directory of Public Relations Firms*. It contains a directory of all United States agencies and their specialties. It also contains a listing of their clients. Conversely, you can look up the client you might like to work with and find out who represents them. The names of the appropriate contacts, vice presidents, and other relevant individuals are listed, as well as phone numbers, addresses, and listings of branch offices. This directory can be obtained in the career center of most libraries. The PRSA in New York City has a large research library that can be used for a fee.

O'Dwyer's also publishes a directory that lists communication officers in major American companies. This directory can give you a feel for the type of communications program a company has—how many people are employed in the department, who is the contact person, and so on.

HOW DO I PREPARE FOR A JOB INTERVIEW?

It is important to do your research before the job interview. Make sure you view the company's Web site and that you have a keen sense of the company's mission and its direction. Try to find out as much as you can about the company—read the annual report, look at brochures, try to learn if they have won any awards, look up the background of the person who is interviewing you: are they in *Who's Who*? Have they received any awards? What is their reputation? Has anyone written about them? You might also want to conduct a Nexis search. This is a computer database search that allows you to view articles written about a person, company, or subject.

The only way to impress a potential employer that you are serious about a communications job is to demonstrate your breath of knowledge about the PR field and the specific knowledge you have about their company—its communications strategy as you see it. That, and establishing good rapport, will likely land you a second interview and possibly even a job offer.

Remember, you goal is to get your foot in the door and to find a job where you can learn as much as possible. Be persistent. One Ph.D. recently got a job with us only after numerous interviews. He maintained that he could learn PR and that he was so motivated that he would start at any salary at any level. He had researched the profession and he had spoken with a number of headhunters in the field, whom he identified through newspaper ads, and he narrowed his search to one agency. With that kind of sales pitch, we had to hire him.

THE PERSONALITY PROFILE

To be successful in communications, it helps to be an extrovert. Type B personalities—the calmer personalities—are better suited for this profession. Remember, you will find yourself in many crisis situations and you will have to maintain calm. On the agency side, there is a crisis at every turn, and the ability to maintain one's cool while attending to the many details of a difficult situation is a personality trait that the job demands.

The profile is slightly different when you are in house. It represents the opportunity to learn a lot about a single company—to fully understand the drug development process, to be a spokesperson for the organization. While you may not have to attend to the volume of potential crisis work, you have to be able to navigate through the corporate structure. Your job and your responsibilities may have a greater likelihood of change as the organization evolves. A drug that is not approved by the FDA can have a severe impact on job viability in the public relations area, which is often viewed as the most expendable department in a company.

To deal with this type of change you need to be flexible and to look for opportunities within an organization where you can add value in crisis situations to help a company rebuild. PR requires a number of skills, many of which cannot be learned. You need to be extroverted, gregarious, personable, decisive, and a good and quick writer, and you must demonstrate confidence and a sense of control. While other skills are helpful, including knowledge of the industry, business acumen, and knowledge of science, personality and persona count enormously.

Unlike lab work, where you often work in isolation, PR is a group-oriented profession. In an agency, there may be many individuals focused on different areas. The "gestalt" requires interaction among the group—brainstorming, sending proposals to the client, and distributing actions among the group. You seldom work in isolation.

A corporate setting may have similar strictures and a hierarchy of approvals may be required before action is taken. In smaller companies, you are able to go directly to the president with a proposal.

DIFFICULTIES AND PLEASURES

"Stressful" is a word that is used frequently to describe public relations—and for good reason. Why is it stressful? To balance so many projects at different levels all at once, and to attend to so many details simultaneously, knowing that one dropped ball can cause incredible chaos, can take its toll.

In an agency, each client thinks they are your only client. Crises always occur at the worst possible time. Chernobyl happened at 1:23 A.M., Three Mile Island occurred just after 4 A.M., and the *Exxon Valdez* accident happened shortly after midnight. A client does not care that you might have five other news releases to get out, all of which are very important to other clients. The Food and Drug Administration doesn't consult with you about when they will approve—or reject—a product.

But if you can manage and if you like a challenge, PR gives you great opportunity for creativity. Unlike the lab where there is a procedure for everything, PR is an open universe. There is no science. There are precedents, but few hard and fast rules. It gives you great latitude and the chance to try new ways of doing things. While there may be certain formats, certain ways of writing, it is in no way as rigorous as science. In fact, in PR creativity is at a premium. The best professionals are the ones who are the most creative.

But what is creativity? Many people imagine that public relations is like advertising. People get together and dream up an ad campaign and brainstorm lots of ideas. In public relations, the ability to have this sort of brainstorming session can be limited. Creativity in public relations is the ability to help maneuver clients out of difficult situations: a drug fails in clinical trials; a CEO resigns; a competitor releases a study that questions the viability of your drug; or, on the positive side, your client just got published in *Nature*, so you need to decide what to do. There is a limited number of options, and you have to consider that you might have to deal with the Securities and Exchange Commission or with FDA regulations. Within these guidelines, however, there is room to deal creatively with various audiences: the general public, investors, physicians, the scientific community, and so on.

WHERE DO I GO TO FIND A PR JOB?

One of the best places to find a job these days is on the Internet. There are lots of jobs listed on Job Trak and on other university postings. Also, some public relations agencies, such as Noonan/Russo, biotech, and pharmaceutical firms, also post jobs on their home pages.

Still, the best way to find a job is to do a little research. Look through the O'Dwyer's listing of PR and communications positions in the library. Do as much research as possible, then send a pitch letter to the firm in which you are interested. Be specific about why you want to work there. Why are you interested in them? Why does your background make you a good fit?

Another way to find a job is through the old network system. Attend some of the investor conferences. Get to know people in the industry. Perhaps you could visit with companies and inquire as to what type of skills they require. Then you could build up your résumé, either through course work or as a volunteer.

Finally, you could read the classifieds in major newspapers such as the *New York Times* and the *Wall Street Journal*. The PRSA local chapters also post jobs in their state newsletters as does O'Dwyer's PR Marketplace. Information on all these publications is available through the PRSA.

And be persistent!

17
· · · · · · · · ·

SALES AND MARKETING:

So You Want to Sell?

· · · · · · · · · · · · · · · · ·

Erin Hall Meade, M.S.
Pacific Northwest Sales Manager, LAS Laboratories

My name is Erin Hall Meade, and I'm the Pacific Northwest Sales Manager for LAS Laboratories, one of the largest environmental testing labs in the nation. I have an honors B.S. degree in medical microbiology, with a minor in chemistry. I made it through all the course work and lab work for a Ph.D. in medical microbiology and infectious diseases, and I am a Registered Microbiologist and a Specialist Microbiologist. Although I sat on the Board of the National Registry of Microbiologists for several years and am a Fellow of the American Academy of Microbiologists, I have never worked as a laboratory microbiologist. I have been in technical sales for the past 13 years, and I *love* it.

How Did I Get Here?

In 1980, I had just passed my qualifying oral exams for my doctorate in microbiology. Ahead of me loomed 6 months of working in a small windowless room, refining my piles of research into an acceptable dissertation. Somehow, I was uninspired by the prospect.

That night, I lost even more of my motivation when I opened the *Journal of Bacteriology* and looked at the postdoctoral positions listed in the back. The salaries were between $8,000 and $11,000 per year, barely living wages, even for an ex-grad student used to living on the financial edge. Two days later, while I was still stewing over my impending fate, a friend called and told me that a small chemistry laboratory in the San Francisco Bay area was looking for a lab manager and paying $18,000 a year. I walked away from the dissertation, and never regretted it.

I spent 1 year with that first company. When it went bankrupt, I jumped to its largest competitor, again as lab manager, for a 50 % increase in salary. I lasted there for 5 years, then was "laid off." (Actually, I was fired because I was in the wrong political camp during a power shift. Oh, well.) I noticed that the *only* people not canned were the salespeople! When I asked why, I was told, "they're too valuable to lose—they are the ones bringing in all the money!" The point was well taken.

I quickly began looking for a sales job, but was repeatedly told, "You're too smart, you'll talk over people's heads; You have no experience; Technical types can't do sales, they're too arrogant, they lack the common touch; and How do you know you can do sales, anyway?" All were valid questions.

I started looking for a company that would train me to do sales, one that specifically advertised "no experience needed." I expected to take a drop in salary at first—and I certainly did. In 1984, I went from $50,000 per year to $35,000 per year, and gave up a company car. But, I did learn to do outside sales, and found that I loved it.

I loved the freedom of making my own hours; I loved the challenge of selling something technical to other techno-weenies like myself; and *I loved winning.* Losing was no great shakes, but I soon learned not to dwell on the losses—they would be wins next time, I decided. I also loved the money; the sales people are often some of the highest paid people in a company. And rightly so—they are the main funnel for cash into the business. Of course, the other employees are equally valuable. But, frankly, they are not perceived as such by the people making most of the salary decisions. The job security is excellent—why would you lay off someone bringing in roughly 10 to 20 times their gross salary every year? (And, by the way, this is a good rule of thumb for a salesperson to calculate what they should be earning—figure you need to sell 10 to 20 times what you make in gross salary, including all benefits and perqs.)

I didn't sell anything high tech in that first sales job; in fact, I sold some pretty disgusting stuff (industrial maintenance chemicals used for exciting applications such as cleaning grease traps for french fry machines at McDonalds). But I got good at it by paying attention, working 14 hours a day, reading all the company literature that no one else reads, and doing a lot of planning and strategizing before I left the house every morning.

So, I learned to do sales, lasted one and a half years, and jumped into a technical sales job in another company working in a technical field—environmental laboratory services. I wanted to sell something interesting and sell to more intellectually challenging clients than those buying cleaning solutions. I immediately jumped back to $50,000, got a company car again, and was selling something high tech, way cool, and was feeling good.

Let's face it—you didn't sit in school for 16 to 24 years to be satisfied selling nontech stuff. That's a commodity market, and doesn't challenge you, nor does it really utilize all those cool resources you've accrued over the years. Selling technically interesting things (or services, or whatever) forces you to use your brain, draw on all the knowledge you've accumulated, and be there all the time.

A technical degree (or two, or three) gives you two distinct advantages over the other salespeople in your field. First, it teaches you to think on your feet, to look at loads of information and identify the salient components needed to answer the clients' questions, and to distinguish between useful and garbage information. Second, most of your competition is hiring sales reps with nontechnical backgrounds, who sell by the "Hey, buddy, how about them Niners?" technique. While this approach can work in certain settings, most clients buying pharmaceuticals, laboratory services, or some other specialized service will be greatly annoyed by it. Few, if any, of my clients want to talk about the big game.

When a client asks me a technical question, I don't have to say, "I don't know; I'll have to ask someone *smart* and get back to you." Instead I say, "Sure, you can pick up phenols in a standard GCMS scan; which ones do you need to see?" This ability to grasp the question and come up with a useful answer saves my client time, and often can lead to my selling even more services than the client originally anticipated buying.

Not only do I use my scientific background, but I also use the ability covertly installed in me during school to think on my feet, not to get distracted by extraneous information, to meet deadlines and budgets *no matter what,* and to make the extra effort to ensure everything is done perfectly.

WHAT DO I DO?

I sell environmental testing services for one of the largest environmental testing laboratories in the nation. My territory includes everything from Fresno, California, to the North Pole, from the Pacific Ocean east to Boise, Idaho. This sounds like a large territory—and it is—but there are only about 7 key cities in it, which is where I focus most of my attention.

I travel a *lot*—about 3 weeks out of 4, on average. I have learned to live out of a suitcase, and I have learned to enjoy it. In fact, after all this time,

I can't imagine opening my eyes every morning to the same scenery. How dull.

I make client calls on both new and existing clients; I write quotes, bids, and proposals; I review contracts prior to sending them to the legal department (so I can spot problems early in the game, and often defuse them before they become deal-stoppers); I help clients sort out problems; and I keep *extensive* records, so if I'm hit by a bus, someone could step in and take over my territory with minimal hassle.

Part of the reason for good record keeping is that the environmental testing industry is driven by state and federal regulations. Few companies would choose voluntarily to spend $50,000 to $500,000 per year just to monitor their environmental messes. Therefore, the federal government has made it a legal requirement to do so.

Often the salesperson is in the middle of a huge web of regulations, requirements, suggestions, and just plain field realities that need to be sorted out, coordinated, and pulled back together to make the project work. Which regulations apply first? How do you meet the needs of multiple regulators, field personnel, accountants, engineers, lab people, and still keep everyone playing nice with each other? The case can be made that this is not the salesperson's problem. Fine, I agree . . . Of course, then nothing gets done, or it gets done wrong, or the client is unhappy, the regulator is unhappy, and you're in trouble. So, let's rethink this. Whose responsibility is it to make sure everything goes well? *Yours.*

And you're handling problems from both sides; your company (those wonderful folks that say to you, "You sell it, we'll take care of making it; doing it; designing it; implementing it.") have been known to let you down ("gee, it was my day off . . . I don't know where the fax went, but I never got it . . . well, we've been *very* busy lately and it must just have slipped through the cracks," and so on. Then you are in the unenviable position of selling the problem to the client ("No, it's not really a problem, it's an *opportunity for personal growth*").

How Much Do I Make?

There are five ways in which sales people can be paid; you can usually get a combination of these.

Salary

This is a flat rate, paid in weekly, fortnightly, or monthly increments; increases only when you get a raise, usually only once or twice a year. Advantages—dependable. You always know from month to month what your financial baseline will be. Salary varies with the job, the company, and their expectations of you. An entry-level salesperson typically will get from $25,000 to $35,000 to start. Someone with around 5 years of relevant experience and a good track record should look for $45,000 to $75,000. A senior sales rep with a good track record of sales should expect from $65,000 to $120,000 annually. If you're in a hot industry (software is big right now), you can make a lot more.

Commission

This is a percentage of the revenue income that you bring into the company through your sales activities. Avoid a "commission-only" position, particularly if you're just starting in a job. Usually it takes several months to establish a client base in a new position and get sales revenues up to a meaningful level. If you receive only commission, you'll be working for very little income (sometimes no income, if the company has a revenue cap that must be exceeded for commission to kick in) until that point.

As you become established, commissions are a good way to go—it gives you a way to determine your own income, based on your selling skills, the amount of time you put into the job, and the real value of the products you're selling.

Commission-only settings are best left to the seasoned sales reps, who have a realistic idea of their sales abilities, an established client base, and can continue to sell successfully month after month. But even as a beginning sales rep, you are bringing value to the company other than revenue. You're advertising for the company that hired you, and you're building client contacts and name recognition for the company that will last long after you are gone. Because the company gains intangibles from your activities, it should pay you some baseline salary that underlies your commission, to keep you going until the commissions start.

Base salary

This is the minimal amount people on a commission plan get from the company, on a regular basis, to handle basic support. This is usually anywhere from $25,000 to $60,000 annually. Commissions add on to this.

Draw

This is an amount of money a company gives you to get you started. Be very careful—*it is only a loan* and will be repaid out of your commissions (when they start), or, if you quit, is often repaid out of your pocket. When a company uses a draw to get you going, you're essentially working for free, as in the commission-only scenario. This is rarely a good idea.

Bonus

This is the happy event that happens on an irregular or regular basis, and is a reward for doing something particularly well—exceeding your sales quota, landing a particularly good account, or helping the entire team have a good sales period. Bonuses are a great incentive—they're a slug of cash all at once, and really perk up your day! Always ask, when you're hired, if your plan includes a bonus and how you can qualify for it.

Perquisites

These are the nifty little "gimmies" that are nontaxable, nonreportable (by you), and make life nicer. Examples include a company car; a gasoline credit card for which the company pays; a cell phone for which the company pays; a laptop or desktop computer that the company buys; an expense account; and so forth (like the right to keep all your frequent flyer miles, which can translate into neat, cheap vacations).

The best part is that Uncle Sam gets little or none of this stuff because it's not really cash, and it's not really given to you—you're just using it to do work for the company. It is important to remember that perqs should never be abused, or they can disappear (often, along with your job). Don't use your car phone to order take-out pizza or use your company credit card to take your mom to dinner. However, you *CAN* use your credit card to take a client to dinner—as long as you discuss business during that time.

Similarly, you cannot drive the company car across the country for a family reunion without reporting this as "income" on your taxes (and, to the company, as "personal use," if that is allowed), but you can use it to drive to another city to visit a client and (happy coincidence!) your best friend, who happens to live in the same city. Be smart, be honest, play by the rules, and this all works out wonderfully.

Extra Option

Some firms, particularly start-ups, can't afford a senior sales rep but truly need one. In this case, they may offer stock or partial ownership in lieu of a juicy salary. This is a judgment call—you're gambling that the company will become hugely successful and you'll make your money back in spades. Go with your gut feeling. If you think this technology is the greatest thing since crunchy peanut butter, then do it. However, if it's a "me, too" technology, is half baked, or otherwise smells a little whiffy, politely decline. Cash in pocket is often better than founder's stock that matures in 7 years (and in which you vest in 5 years); you're betting today's earnings against stock performance 5 to 10 years in the future.

HOW DO YOU PREPARE FOR THIS SORT OF CAREER?

The big catch-22 is that you can't get a sales job until you have sales experience, but you can't get sales experience until they hire you to do sales, . . . Right? Wrong. You can take a job as an in-house sales representative (entry level position, usually involving phone work), do that for a year or so, and learn sales that way—plus show the company what you can do. Or, you can get a job in client support/customer service, supporting the sales department, and learn that way. Or, look for a company that's willing to train you, as long as you swear to work for them for a certain amount of time once training is complete. Any of these approaches work well. My estimate is that you'll spend from 1 to 2 years in inside sales, 1 to 2 years in low-level outside sales, and then you're totally out on your own—a terrific place to be.

Is it necessary to know something about an industry to sell for it? Yes . . . and no. Knowing something about the industry can help you get a foot in the door, can give you an advantage over other job applicants, can help you manage your sales territory for the first few months, and can help you target the appropriate accounts right off the bat. However, you'll really do 99% of your learning on the job. I had *zero* knowledge about each industry when I first started in them. I paid attention, sold my sales and/or management skills over my knowledge of the industry, and learned fast. It worked for me.

The typical promotion pathway often looks like this:

Inside sales (i.e., phone sales) → outside sales, junior grade → outside sales → product manager → sales manager → director of sales → vice president of sales.

Note that inside phone sales *is not* telemarketing. You do not annoy people by calling them at dinnertime; you sit in the company office and answer phone calls from new or existing clients who need information or wish to place an order. Inside sales is the bottom of the ladder—you have a tough job, you have to show up in the office, dressed appropriately, and on time on a daily basis; people yell at you over the phone; you never get to meet anyone face-to-face, which makes it much harder to develop a relationship with your client, and the money is not great. The good news is that it doesn't last forever. Think of it as your internship.

Outside sales, junior grade, means that you often travel with a senior person, who shows you the ropes. This allows you to learn a lot in a short time. While the money isn't great, it's better than inside sales.

Product manager may or may not be a promotion. When you're a product manager, you have the responsibility of building name recognition and a client base for one particular product, but you don't sell anything else, which can be limiting, and can cut you off from your former client base. However, if you make a new product a huge success, you get all sorts of money and recognition within the company, which is good for promotability.

Becoming sales director is the holy grail for some salespeople, while others prefer to lay low and live the good life as a field salesperson. Personality usually determines whether you will want to go to this stage. Being a director means you spend your time managing salespeople, who are also known as "cowboys" because of their cussedness and independence. After being a manager for 5 years, I gratefully dropped back into being a salesperson—now the only person I have to manage is myself. And, contrary to common belief, being the manager does not necessarily mean more money—just more aggravation.

Becoming vice president of sales means you take the headaches of being Sales Director, multiply them 10-fold, then add corporate politics on top. Personally, I have no interest in this position. However, the money is really good—expect to make $150,000 to $250,000 annually, with performance bonuses on top of that. If you're in a hot company, you can do even better .

CAREER MOBILITY

If you get tired of sales and decide to switch careers or companies, you should know that a good sales rep can go just about anywhere, given the proper network. Selling is an innate talent, as well as a trainable skill, and not everyone can do it. This makes you valuable. You can switch companies easily, switch industries with a bit more effort, and move to a very senior position within a corporation with diligence. A majority of CEOs in Fortune 500 companies formerly were sales reps, according to *Forbes Magazine*.

The key here is *who you know*—networking is everything. The higher up a sales position is, the less likely it is to be advertised, and the more likely it is to be filled by connections, networking, and word of mouth. Never overlook headhunters (kindly known as "recruiters" or "search consultants")—some of the best jobs are found through them, so they can winnow through the chaff looking for the truly golden individual for a particular opportunity. *Always* be nice to headhunters—they are good nodes in your network. When they call you, be polite and listen, even if you don't think you are interested. If it's not the right job for you, say so—and then suggest someone else you know who might be a better fit. The headhunters don't forget favors like this. They also call you more often, and thus you are more likely to hear about a truly nifty job.

What about Travel?

Sales absolutely requires an enormous amount of travel. There is a direct correlation between the sales revenues I bring in and the amount of time I am sitting in front of a client. Phone calls, e-mail, and even video conferencing do not bring in the same results. Just the idea that you thought enough of someone to get on an airplane to come see them warms the cockles of even the hardest heart—especially if you do it regularly.

I travel about 2 to 3 weeks every month, hitting all 7 of the major cities in my sales territory (San Jose, CA; San Francisco, CA; East Bay, CA; Sacramento, CA; Seattle, WA; Richland, WA; and Anchorage, AK). I log about 100,000 airplane miles per year. You really do get used to it. In fact, there are some tricks to make it downright enjoyable.

For example, I find one hotel in every town, and I always stay there. I get to know the staff, the best rooms, the concierges, and (especially) the reservations manager. Whenever I go to that city, I call the reservations manager and tell him or her that I'm coming back to their hotel for a few days. I get the best rooms, for the lowest rates, and people are always happy to see me. In return, I give them all my business in that city; I *always* write a thank-you card to the reservations manager after I leave, and I try to remember key people's names. People *love* to hear their name—it is the loveliest sound in the world to them. (See, I'm always selling!)

Another tip—buy a good map of a city (*Not* a rental car map!), and keep it in a file where you keep all the other pertinent information on that city, including client locations, dry cleaners, hotels, drug stores, and so on. Then, when you get ready to visit that city, just grab the file and throw it in your briefcase. You will become so familiar with the city that, after a few trips, it feels just like home. I learned to like Anchorage so much that I'm moving there.

Try to fly with only one airline, whenever you can. You not only rack up miles (leading to preferential treatment and first class upgrades), you also get to know the gate people—which can come in very handy when a flight is oversold, or you have excess luggage, or you're running late, or for any number of other little problems that life may throw your way.

Learn to pack a suitcase properly. Everyone knows how to pack, right? Wrong. For 5 days out, a woman will need 2 jackets, 5 blouses/shirts, and one dark skirt or pants. Period. One pair of jeans for night, to wear with the same blouse you wore during the day. One pair of shoes—low heels or flats—that match the skirt. One purse to match the shoes. One nightgown. I also bring one pair of running shoes, just for walking around after work.

I leave the rest up to you, but remember—you're probably not going to run into Prince Charles, so don't bring anything extra, it just adds weight. I can go out for 5 days with one small carry-on suitcase. I toured mainland China for 4 weeks with carry-on luggage, and did just fine. There are department stores everywhere, and it takes 5 minutes to pick up an extra pair of panty hose or clean socks. And 24-hour dry cleaners exist, so a coffee stain on a jacket is no big deal—just switch to jacket #2 for a day.

WHAT COMPANIES HAVE SALES PEOPLE?

Easy—*all* companies have sales people. Some companies have different names for them—account representative, business development, marketing, and so on. Any company that sells a thing, an idea, or a service has to have someone to sell it for them. And it might as well be you!

What percentage of a company is composed of salespeople? This can vary widely, depending on what they sell. My lab has 75 to 90 employees, and we have 4 salespeople. Between the four of us, we bring in $9 million to $14 million a year—enough to keep those 90 people employed. If a company is just a "shell"(meaning they're just a middleman, adding cost but relatively little other value to someone else's product other than a sales force), the percentage can be as high as 90%. However, watch out for this type of company—they usually don't take good care of their salespeople and their clients can often cut them out and go directly to the source of the goods or service, thus cutting out the mark-up.

Also, watch out for companies that sell commodity items—boring stuff than *anyone* who can walk and breathe simultaneously can sell. This means things like office supplies, copiers, telephone systems, encyclopedias, maintenance chemicals, paper supplies, restaurant supplies, all that sort of stuff. If it doesn't take any special skills, intelligence, or talent to sell the product, you're in stiff competition with the 23-year-old buxom blonde who giggles and snaps her gum, but is cute as hell—and she's going to get that sale almost every time. It sounds trite, but it's true. Try to sell something that takes brains to understand and explain—there's less competition, you have better job security, and it's a lot more challenging intellectually. Plus, you become more valuable to the company as you get older, not less valuable.

What Skills Are Necessary?

It is critical that you have the ability to meet people and become comfortable with them easily. You also need the ability to talk to strangers, and an ability to accept criticism and rejection without taking it personally. Just because they don't buy your product doesn't mean you're a bad person, they just don't buy from you. No big deal, just go find someone else who *will* buy from you.

You will need to be completely self-motivated, especially in outside sales. No one from your company sees you regularly, no one tells you to get out of bed and make an 8:00 A.M. appointment, no one watches to see what time you quit in the evening—it's all entirely up to you. This can be a fatal trap for someone who is not internally motivated, because while no one is watching you, they are all watching your numbers. If your sales fall below average, you're soon on the street again—with no reference for your next job.

You will also need to be exceptionally well organized, as you will keep your own files. I spend one day per week just on paperwork—filing quotes I've written, responding to requests for information, tracking procurements and projects, keeping up with time cards, expense reports, mileage logs, and tracking customer problems. When I'm out of town for 2 to 3 weeks, this means spending the first three days I am home just playing paperwork catch-up.

You need to have good time management skills, too. You have to get everything done in a reasonable amount of time, or else your job will engulf you and eat up your entire life. Sales takes a lot of time—I work an average of 10 hours a day—but it also gives you the ability to manage your time however you want to. I may play hooky one afternoon and go climb a glacier when I am in Alaska, but I'll work until 11:30 the next evening writing a proposal that is due the following day. There are days I work 4 hours, then quit and lie in the sun; but there are many more days when I work until after dinnertime.

You must be able to write easily and well, as you will be doing a great deal of it—proposals, bids, reports, letters, and so on. If writing is a struggle for you, sales is not your field—you must leave a clear paper trail behind you, for both legal and professional reasons.

Being a confident, aggressive person is a plus. A shrinking violet has a hard time in sales. While confidence comes with time and knowledge of your product and your company, the inherent ability to walk up to a stranger and say, "Hi, I'm Erin Hall Meade, and I work for LAS Laboratories. Can I talk to you for 5 minutes?" is mandatory. Never be embarrassed—if things don't work out, you just walk away and they'll forget you in no time.

A lot of people say good looks are important in outside sales. While it is true the everyone likes to look at handsome people, this is *not* a job requirement. It is more important that you have a pleasant personality, that you are well-groomed, and that you are confident (does this sound like something you heard in your high school health class?). A well-dressed, personable person will always come out ahead of a gorgeous airhead hunk (unless you're selling copiers or carpet shampoo, and why would you want to do that?).

There are a few people in my field (with other companies) that have minimal technical training. I have no difficulty selling against them, even though they may be younger, better-looking, thinner, and so on. When my client says, "How can I do this job for the least money, getting the best data?" the airhead says, "I'll call my office and get a technical person to call you back." I say, "Let me show you how to optimize this project, saving both time and money." My success rate is higher than average because I know my field and the technology, not because I am the best-looking person in the field. (Which, of course, I am. Ha!)

What Does a Typical Day Look Like?

I have two types of days: on the road, and in the office.

In the Office

Right now, my office is in my home, which works out very well for me. At about 7:15 or 7:30 A.M., I check my phone messages on one line, while checking my e-mail on another. While I am doing this, I also pull my faxes from the night before off the facsimile machine. I take 15 to 30 minutes to note the things I need to do that day, with A, B, or C priorities. I also check to see what I didn't finish the day before and add it to this list, again with priorities included.

At about 8:00, I roll my telephone off the night line so that it rings instead of going to voicemail. I don't do this earlier because otherwise I'd never get any planning done first. I continue to answer all voice messages, e-mail, and faxes. Then, I begin calling clients to make future appointments, resolve problems, follow up on proposals and quotes, catch up on local industry gossip, and reinforce relationships with key clients.

After this is done I write reports, fill out time cards, car logs, expense reports, and other paperwork. During the entire day, I am handling phone calls from clients, lab people, other sales reps, government regulators, and so on. I usually do my filing just before I quit for the night, which is between 4:30 and 7:30 P.M.

On the Road

I never go on a road trip without having at least 50% of my sales calls already scheduled with clients. I use a portable cell phone to make the rest of them while I'm driving between sales calls. I write reports, answer e-mail, and get faxes in my hotel at night. I check my voicemail by cell phone from 10 to 15 times each day, since I don't like to let more than an hour elapse between receiving a message and returning the call. I often eat dinner in my room, so I can catch up on paperwork, faxes, and so on. I get a lot of reading done, too.

WHAT ARE THE PROS AND CONS OF THE JOB?

I *love* the freedom, and the fun, and the adrenaline rushes when I win the sale. I love the travel (especially on someone else's nickel), I love the challenge, I love meeting new people, seeing new places, and learning new things daily. I am a true adrenaline junkie—I love the ups and downs of sales. And, best of all, my job is just plain *fun* much of the time. Gee, and they pay me lots for this! Wow!

I *don't* like having no control over things that affect my clients, especially problems in my company's lab that impact its ability to get the work done well and on time. I can't sell in the field and direct things in the laboratory at the same time. Sometimes I feel like a mushroom—the lab keeps me in the dark and feeds me manure. I find out about a problem with a client often after the client does—I hate this. Still, it happens. I have to deal with it, and do damage control wherever possible. This means that I have to maintain a decent relationship with the people at my company in order to keep them responsive to my concerns about my clients. I often have to convince them to do my clients' work before they do someone else's work. In that way, the people inside my company are also my "clients." I have to sell them on the idea of doing my outside clients' work first.

It was easy for me to get used to managing my own time; after all, that's what you do in grad school. No one checks to see if you're going to class, but you'd better do well on the exams or you're sunk. I also had no difficulty getting used to making on-the-spot decisions, sometimes involving hundreds of thousands of dollars. Once I feel I have the technical knowledge to make the right decision, I just do what seems best—and, most of the time, I'm right. When I'm wrong, I face it, and it's okay. I've never been fired for making a bad decision, but you can certainly get fired for not making a decision at all.

How Often Do I Have a Hair-on-Fire Day?

Fairly often, maybe 25% of the time. As I mentioned earlier, I am an adrenaline junkie, so I go for the highs and lows. I *love* days when I have 20 hours of things to do, and only 15 hours in which to do them—this is when I'm at my most creative, productive, and stressed out. Fifteen hours flies by in minutes—I don't eat, I lose track of time, and I love it. If you're not into pressure and stress, you may not like sales. But I always feel something is happening!

I am a perfectionist. I love to win. I want my clients to know they can always count on me to be their advocate in the lab. I want them to see me as a technical resource for them, and I want them to know that I will always advise them about the best way to spend their money, even if it's not with my company on a particular job. In fact, one of the best ways to build credibility with a long-term client is to tell them how to spend less, or how to spend it with one of your competitors. As long as this is truly the best thing for them at the time, they'll remember you as the one who told them the truth and they'll always come back.

Finally, think technically! You have had tons of education, you have had tons of training to help you gather and sort through data and synthesize solutions from mountains of information, and you have been trained to *think on your feet.* This is something other people don't have, so let it work for you. And have fun.

18

EXECUTIVE SEARCH:

Looking for Talent in All the Right Places

• • • • • • • • • • • • • • • •

Bente Hansen, Ph.D.
Executive Vice President and Managing Director of Biotechnology,
DHR International

Although I pursued my advanced degree to prepare for teaching anatomy and physiology in the university, I am now a headhunter—no, not in deepest Africa but in executive search.

How did I get from the classroom to the boardroom? My career has taken many turns and has included a few high-risk entrepreneurial positions, but I have tried to maximize my experiences to enhance my skill set. Experiences that include business development, marketing and sales, writing business plans, writing SBIR grants, starting my own company (Medical Impact), and running a preventive medicine center have helped me prepare for a future to which I am committed and that I am passionate about—executive search.

My transition was not a direct one. I found out about this industry first from people working for executive search firms who contacted me regarding opportunities that they had available. Pay attention if you receive such calls! They are directed to you because the recruiter has researched you as a possible candidate for a prospective job for which they are recruiting or as a source of recommendations.

I was curious about the executive search industry because it seemed like a business with great growth potential, and I started asking questions of the recruiters contacting me. Most search professionals are personable and polite on the telephone. My next step was to meet with someone who was successful in the industry so that I could learn about and better appreciate the day-to-day schedule of an executive recruiter. I was introduced to a highly successful recruiter who owned her own practice and who has made a name for herself in the recruitment of engineers. She was kind enough to give me a tremendous amount of information and support.

I was truly at a crossroads in my career at the time. While completing a 9-month consulting contract, I began to identify which career opportunity would best fit my skills and would best suit my long-term goals. I realized that I was most comfortable in a position where I could use my scientific training and my years of management experience. I determined that my strengths were my knowledge of the biomedical industries and of business development.

My preliminary discovery was that the recruiting industry matched my skills and interests. I did extensive due diligence on the industry, and found that there were few books or related articles on this industry. With *effort,* I obtained some career books including articles written years ago. The book, *Career Makers*, included brief biographical summaries of top national recruiters that gave tremendous insight into the qualities that made these folks so successful.

WHAT I DO NOW

My current job at DHR International can be divided into two major responsibilities:

- To obtain a search contract with a company for the recruitment of senior management on a retained basis;
- To fulfill the contract and locate the individual who will best fit the position specification and corporate culture.

Working as a consultant with DHR while part of a team with specific corporate guidelines enables me to have the flexibility to manage our time and resources as I best see fit.

David Hoffman, nationally recognized as one of the country's top recruiters, established DHR International in January 1989. Since that time, DHR has grown significantly, with offices in 31 cities throughout the United States and with 16 international locations. DHR has been ranked the fastest growing executive search firm in the country, and is the largest firm in the nation in terms of office locations and geographical coverage. The principals of DHR are experienced professionals who have been engaged to conduct executive searches for many of the leading companies in the world.

Our search philosophy and methodology are unique within the executive search industry and have earned our company an excellent reputation. The foundation and fabric of DHR is based upon personal service and dedication to quality, which is evident in DHR's thorough researching capability, the teamwork brought to bear on every assignment, and the timeliness with which we complete our assignments. DHR promises to present a completed list of qualified candidates to the client within 20 working days. This short turnaround is an important value-added dimension of our service.

Responsibilities and Attributes

The responsibility of an executive search consultant is to bring the most qualified candidate for a given position together with the client company. To do this, the headhunter must have a thorough understanding of the client company, its technology, its management structure, its vision, and most importantly, its corporate culture.

My responsibility as a retained executive search consultant can be divided into two distinctive and separate components: (1) business development—gaining new corporate clients for whom I will conduct a search; (2) completing the search assignment once I have secured the contract.

These two roles are very different. One is a marketing and sales component that includes presentations to potential clients and a need to develop a wide and varied contact base. The actual search entails project management skills, and a thorough understanding of the company. The following is a breakdown of the traits needed for both of these roles.

The characteristics needed for marketing/sales/business development include:

- An ability to convince others;
- An ability to identify strengths;
- An involvement in community and professional activities;
- A willingness and eagerness to meet new people;
- Flexibility regarding income stream.

The characteristics needed for completing the job, i.e. finding and attracting the successful candidate, include:

- Knowledge of the industry and good contacts;
- An understanding of management's needs and desires;
- An interest in others;
- Being a self-starter;
- Being flexible with creative thinking;
- Being service-oriented;
- Being analytical and detail-oriented;
- An ability to interpret culture and values of the client organization;
- An ability to lead and direct researchers;
- Patience with changes that need to be made by the client;
- Knowledge of job specifications within your industry;
- Hands-on experience with the personal qualities that are important to success;
- An ability to prioritize and differentiate among highly qualified candidates;
- An ability to summarize search stages to a board of directors.

There are key personality traits that are necessary to be successful in the executive search field. It is very important that you enjoy meeting new people, and that you have the ability to maintain your existing network of contacts. This network is critical to identifying strong candidates for your clients, and for finding new clients. If you are shy, or have difficulty asking questions of people you don't know well, this may not be the job for you.

It is important to be a self-starter—in most cases, nobody is going to call you with the candidates, you are going to have to be proactive in tracking them down. You need to have the interest and curiosity to learn about new technologies and new companies in the industry you serve. Otherwise, you won't be able to help your clients as thoroughly, because you won't really understand their business and the issues that drive that business.

You have to be proficient at project management and you have to be able to track and monitor multiple searches while meeting key deadlines. In most cases, you will be juggling several client projects simultaneously, and all of those clients will expect you to stay on top of their project. Every delay they face in bringing the best person on board impacts their company's ability to meet its own milestones. This means that you must be flexible in your work hours to accomodate clients and candidates on the East and West coasts, and the fact that often you can speak with candidates only outside of regular work hours.

It is also important to work with a "win–win" mindset, to be convinced that both the company and the candidate will be enhanced by their relationship.

Skills Needed:

- The objective/analytical ability to evaluate needs of clients;
- You must be secure and comfortable in making presentations to upper management and boards of directors;
- You must be a quick learner with an ability to comprehend and communicate information about numerous companies;
- You must have an interest in meeting new people;
- An ability to write and summarize technical information in layman's terms;
- An ability to manage one's time as an independent contractor;
- An ability to juggle a multitude of projects that are at different stages of development;
- An interest and ability to communicate well in phone conversations;
- You must be very interested in providing service to the client;
- An ability to analyze the culture of a company in order to find candidates who are the best fit;
- You must be able to bring out information from potential candidates because résumés don't always say all that you need;
- Good project management of time and resources;
- An ability to work with many different types of people from the CEO to the director of biology to the vice president of manufacturing;
- You must be service-oriented;
- You must have an interest in complete follow-up.

A Typical Day

You can expect to keep long hours, and to spend large amounts of time on the telephone either speaking with people or, more likely, playing phone tag with their voice mail. Here is a typical day for me:

> 6 A.M.: a 2-hour conference call with the vice president of human resources at a client in Texas to review candidates for the manager of information systems position; check voice mail/e-mail and respond if there is an urgent issue.

> 9 A.M.–12:00 noon: Check references for a vice president of corporate communications candidate for a biotech client; make contact with some of the references, leave voice-mail messages for return calls. Prepare for candidate visit for vice president of research and for candidate visit for vice president of research and development at a biotech firm;

suggest possible agenda for visit with client to follow up on their most recent hire (vice president of clinical research). Contact a real estate agent who will meet with top candidate to discuss relocation issues. Respond to multiple calls from individuals looking for job opportunities. Follow up with committee members regarding an upcoming workshop. Meet with DHR researcher to organize and strategize plans to locate candidates for a vice president of manufacturing position.

12:00–1:30 P.M.: Marketing lunch with CEO of potential client to convince her to use DHR rather than another search firm; mention my relevant industry experience.

2–3 P.M.: Meet with client to get additional information about software company for writing up a position specification. Meet with job seeker to review their résumé; assist with career advancement program.

3–5 P.M.: Committee meeting for industry-related networking group; American Lung Association Board conference call; business development: send letters/do follow-up calls. Return phone calls, e-mails.

5–5:30 P.M.: MIT Enterprise Forum—marketing meeting.

5:30–8 P.M.: MIT Enterprise Forum—networking and program. Presentation.

8 P.M.: Check voice-mail and prepare for next day's activities.

Daily Frustration Level

As with any job, there are a multitude of things that can go wrong in a job search assignment. You may be working on five to eight search assignments simultaneously, which are at different levels of completion. I will highlight a few frustrating scenarios from my experience.

Scenario I:

You must deal wih potential candidates for a client's position, who say that they are interested in the position you have described, yet do not send their résumé as promised. This makes it difficult to verify their intcrest, and impossible to check their background and experience. Usually, there is time urgency on a search assignment and without the CV, you cannot include them in your list of potential candidates.

Scenario II:

Your top candidate rejects the client's offer after you have negotiated for weeks and you thought you had come to an agreement. The reference checking usually is time-consuming and to come to terms regarding the entire package takes creativity and effort. If the top candidate falls out at this stage, often you may have lost the second and third choices to other positions because of the time delay. It is likely that your top choice has developed cold feet at leaving a known environment and moving on to something new and different, but it is probably because he or she was offered a higher compensation package to stay at the current position. Often, one must start the search again.

Scenario III:

The candidate you brought in to the company cannot get along with the CEO or COO. DHR has a 2-year guarantee for each search assignment. This means that if the individual does not stay in the job for at least 2 years, the search consultant must fulfill the responsibility and do another search at no charge to the client. This can happen regardless of how diligent and careful you are in the match-making process.

One of the most important elements of the search is to match personalities and corporate cultures. As a search consultant, you have the responsibility to assess the corporate culture and to bring in the key individuals who will fit into such an environment. Through no fault of your own, there may be pressures or stress that can cause individuals to react differently from your assessment. This may have a negative influence on the new candidate or on the corporate decision maker.

Scenario IV:

In completing your degree check, you find that the candidate does not have the degrees as stated on the résumé. This degree check is done early in the search, so you haven't lost much time. This situation does happen and it is an important and necessary step in reference checking.

Scenario V:

After presenting three top candidates who match the position specifications perfectly, the top management of the company realizes that the position specifications need to be modified. This situation occurs because the management, during the course of the search assignment, does an in-depth assessment (often for the first time), of the existing skills, of projected growth and direction, and of the financial resources of the company. When this occurs, the job specifications are modified and there is generally a clearer vision of the needed skills for this position—but it does entail generating a new list of candidates.

Scenario VI:

After extensive interviews with the top candidate, she backs out because her family decides that relocating from the East Coast to the West Coast won't work out. Relocation is a major variable and always has to be discussed extensively, since it is a life-style change. Often it is a difficult process for the entire family to move, especially when they have school-age children. Sometimes the applicant must decide to be away from family until an appropriate moving opportunity presents itself.

Compensation, Experience, and Advancement

Before I discuss compensation, it is important to discuss the difference between retained and contingency recruiting. DHR is a retained search firm, which means that the hiring company retains a search firm on a contractual basis to do the search, regardless of whether a hire is actually made. A contingent search firm gets paid on a per-hire basis. The fees to the search firm typically are related in some way to the annual salary of the position being filled at the client company.

According to the *Executive Recruiter News*, there are more than 1859 executive search firms in the United States; 931 are retained firms, and 928 are contingency firms. Combined, the search firms gross more that $2.5 billion in annual revenues and are staffed by some 11,724 recruiters.

Advancement within an executive search firm generally is tied to the amount of money that you bring into the firm and the extent to which you are perceived as a leader in your industry. Opportunities can range from positions such as the managing director of the firm's office, the managing director of a specific industry such as telecommunications, biotechnology, and so on, or director of research on a national or an international basis. Along with advancement comes greater visibility, and hopefully, more search assignments.

The compensation in the retained search industry is related to the number of searches one completes and the compensation package that accompanies the search assignment. It is not only the number of completed search assignments that matters, but the salary that accompanies the position for which you are recruiting. In other words, it is better to be recruiting CEOs than lab techs, from a compensation point of view.

At the top search firms, the usual fee paid by the client company is one-third of the estimated first year's total cash compensation of the individual who is hired. This fee is divided into three installments—the initial retainer is paid at the start of the search, then the remaining two service charges are billed in 30 and 60 days, respectively. Often the final payment is made when the successful candidate has been identified and has a signed contract.

The search consultant earns a percentage, usually between 30% and 50% of the amount which is paid by the client to the search firm. Of course, if you own your own firm, you keep 100%! Most search consultants are paid on commission—they are paid when they make a placement—which means that the start-up period may be tedious and the early earnings may be slim. With persistence, experience, relationship building, and successful search completions, one will build a client base that can lead to a solid six-figure income.

Generally, the search consultant in a contingency-based practice has a market niche and has acquired some loyal clients. The fee is still paid after the candidate is hired, but the firm has a track record with this search consultant and knows the talent this person can find. Advancement for the contingency practice is based on getting into an industry niche that is in demand and growing, and on specializing in a certain market such as engineering or focusing on a specialized industry such as telecommunications.

The Rewards of Being in the Executive Search Business

Before I entered this industry, I did a comprehensive skill assessment and thoroughly researched the field of executive search. Like many others, I have had the challenge of job hunting. I had some difficult and unpleasant experiences in the process, and I vowed that I would help people in this situation when I was able.

This brings me to one of the rewards of my job—assisting those looking for a job when I can. In most cases, the people I help are not candidates for the positions that I am filling for a client, so I can't offer a position. I usually can be helpful in revising a résumé to make it more marketable, targeting companies likely to value the candidate, offering contact names who may be of assistance, or suggesting an industry conference that would be beneficial in networking.

Besides the monetary compensation, an executive search consultant must derive pleasure from serving as matchmaker between corporations and qualified candidates. Each person who is added to the management team makes a tremendous difference in the productivity of the corporation. This desire to bring together the right people with the right company for their mutual benefit is critical to being a successful and motivated headhunter. It is rewarding to see individuals find a position that matches their skills and their goals, and allows them to be successful.

Another reward is learning about a broad range of corporations and technologies. To sell the opportunity to potential candidates, you first must yourself understand the corporate mission, products or science, and culture. You will spend a great deal of time on location at the client company, listening to and talking with people who hold various positions within the company, and gaining a real understanding of how that organization functions, its strengths and weaknesses.

This vocation provides you with the opportunity to meet and work with other service providers who often help each other in identifying new clients. Service providers, including attorneys, accountants, public relations groups, and bankers, are all are needed by growing companies to help guide the management team and the board of directors.

In a rapidly changing business environment, search firms are taking on an expanding role within the companies they serve. The search consultant brings added value by assisting the company with other business issues, such as suggesting outside contractors, and assisting with future skill identification and compensation information.

How to Enter This Industry

The job market is changing as our world is changing. The person with a *curriculum vitae* showing a job change every 2 years is no longer considered a high-risk hire, as long as the move was planned and it demonstrates progressively increased responsibility. Today's businesses are looking for individuals with myriad experiences, who can rise to the challenge of ever-growing companies.

How do you orchestrate these "planned" moves? Look at yourself as a small company, "Me, Inc." Take charge of your career and create a board of directors (people who are concerned about your career and who can be a sounding board to guide you). Identify five to seven individuals whom you respect and who respect you as well. Choose persons who are successful, forward thinking, well connected, and willing to help. College advisors, work supervisors, and mentors are good choices. Select those who will give you a variety of perspectives and will encourage you to grow.

With your knowledge of science, you are ahead of the pack in this technology-driven society. As you develop your career plan, build relationships early in your career and nurture them by keeping in touch with those you meet. You will have opportunities to do favors and you will formulate your circle of business contacts. This takes time and effort but the rewards will be great.

There are a number of ways by which you can enter this industry. The usual way has been to gain experience in recruiting and in interviewing via the human resources profession. This is certainly one way to learn about screening candidates, conducting telephone and face-to-face interviews, evaluating resumes—skills needed in this profession.

The other avenue is to work up through the management ranks and to gain experience with a number of different companies or job responsibilities. A necessary component of being a supervisor or a manager is to recruit and evaluate new staff. In this way, one can see firsthand the impact a new hire can have on a company.

Another way to be introduced to the industry is to ask a recruiter who happens to call on you. Be helpful and polite, because you never know when you may be turning to them for an opportunity. I can guarantee that your name will be remembered if you have been helpful.

Finally, there is a list of search/recruiting firms for your city available through the Chamber of Commerce or through a business directory. Contact them directly or ask for an introduction. Some search firms, such as DHR International, focus on attracting leaders in their field, so industry experience is important to them.

Each firm has its own particular qualifications. Some look for industry experience, others look for relevant experience in managing and hiring key people. But the most valuable asset you bring to any company is your contact network. At an early stage of your career, build up those contacts—keep in touch, do favors, keep in touch, attend meetings, speak at meetings, and keep in touch!

PERSONAL COMMENTS

The job market is continually changing and evolving. In the small, entrepreneurial companies with which I often work, there are some key qualities that employers seek: a broad-based set of experiences and skills, project management, and a passion about the company and the technology. This is often an opportunity for people to gain experience in a wider range of areas, since employees do a little bit of everything in a small start-up.

I enjoy my work at DHR. David Hoffman, the founder and CEO, built a team of search consultants by attracting executives from industry who brought with them experience in their chosen fields and credibility with other managers in this industry. In other words, the strength that I brought to DHR was my experience and credibility in the bioscience industry. I am continually learning about this industry, but I have an obvious advantage because I have an advanced degree in science and I have experience in a variety of management roles in the biotechnology, biomedical, and medical fields. Executive recruiting is a great alternative career for people with a science degree.

19

•••••••••

THE GROWTH OF
A MANAGER:

From Pure Research to Policy Administration

••••••••••••••••

P. W. "Bo" Hammer, Ph.D.
American Institute of Physics

What is a Ph.D. and what is it good for? To zeroth order, as physicists would say, the Ph.D. is a scholarly degree that trains people in scientific inquiry and research. One also learns to write and speak in the course of presenting and defending his or her findings. Occasionally, some teaching experience is thrown in. Hence, the Ph.D. is designed to qualify one to work in academia as a scholar doing research and preparing the next generation of Ph.D.'s.

Given this, many are surprised to learn that fewer than 8% of those with a bachelor's degree in physics actually end up filling this Ph.D.-to-academia career niche. The other 92% land in a range of professions as varied as the individuals holding the physics sheepskin. I am one of that 92 percent. Although I earned my Ph.D. at a prestigious university, studying under a respected advisor and then proceeding to a postdoctoral fellowship in one of the most competitive groups in my field, I am now working as Assistant Manager in the Education Division at the American Institute of Physics.

WHAT IS ORGANIZATION MANAGEMENT?

What qualifies me for my job in organization management? First and foremost, according to the job description, was that the applicant must have a Ph.D. in physics. That was simple, but what is it about managing a group that runs physics education programs that makes the Ph.D. necessary?

The primary answer has to do with communication. Besides training for research, the Ph.D. is also a socialization process whereby one becomes a member of a community with common traditions, mores, and ways of communicating. To be effective within the physics community and to have credibility, it is beneficial for one to be a part of the physics culture. It is not necessary, of course, but it helps substantially to be able to communicate and empathize with physicists as a physicist.

The second item on the job description stated that the applicant must have considerable experience as a physics educator. This was problematic for me, because the extent of my teaching background was as a teaching assistant in graduate school. To get the job, I would therefore have to persuade the people doing the hiring that I would bring other qualities and experiences to the Division that would mitigate my lack of teaching experience.

At this point it may be instructive to elaborate on what the AIP Education Division does. Our primary job is to run the Society of Physics Students (SPS) and Sigma Pi Sigma. SPS is a professional society, primarily for undergraduate physics students, with about 6,000 members. Sigma Pi Sigma is the physics honor society, with about 35,000 members. Running these two organizations requires managing staff and coordinating the range of services provided for our members.

In addition to administering these two organizations, our division publishes a hands-on science magazine, an electronic physics education newsletter, and we are initiating new programs to serve physics students and educators. Thus, to manage this group effectively requires experience in administration, management, organization, and also vision and ideas about the Division's future and its role within the physics community and society. It was my professional experience after my postdoc that added the real value to my Ph.D. and qualified me for my job.

HOW DID THIS HAPPEN TO ME?

Near the end of my postdoc, I decided that I should expand my career options by getting experience in science policy. This decision was motivated by my apprehension about increasing competition in the academic job market, and by my own interests in policy and the role of science in society.

While reading *Physics Today*, I learned about the Congressional Science Fellows program and the physicists who became fellows. Over the next couple of years, I became increasingly intrigued with the idea of applying to the program. Discussions on the Young Scientists Network and my own perceptions of the world in 1993 catalyzed my decision to go for it. On my birthday that year, I got the call telling me that I had been selected by the American Physical Society to be one of their two fellows for the 1993–94 class.

Congressional Fellows are sponsored by a large number of professional societies. The program is designed to bring scientific expertise into the legislative process, while simultaneously exposing the science community to the intricacies of how policy is made at the federal level. Fellows spend a year on the staff of either a Member of Congress or a Congressional Committee.

Ostensibly, Fellows are brought in to be the staff science expert. In reality, they become integrated into the staff mix and work on a variety of issues, few of which have science content. I chose to work for the House Science Committee, so my immersion in science policy was greater than that many of the others in my class. My year on Capitol Hill left me poised to make a career transition into science policy and management.

My next move was into another postdoctoral-type position, only this time it was not in science or policy, but in management. Again, an ad in *Physics Today* caught my eye. This job was dubbed "Physics Management Fellow," a temporary position designed to train a young physicist in management while serving as an assistant to the Executive Director of AIP. This job was a natural follow-on to the Congressional Fellowship and it honed my writing, speaking, management, and negotiating skills.

More importantly, being a Physics Management Fellow was an opportunity to positively impact an organization in which there were opportunities for professional growth into permanent positions. Or, it could be a good launching point into a permanent policy or management position elsewhere.

When the Education job opened up 2 years later, I was in a very good position to apply. My main challenge was to persuade AIP that the diversity of my background would compensate for my lack of direct experience as an educator. My two years in the Director's Office allowed me to gain the confidence of AIP management and demonstrate my abilities. Had I applied from the outside, I probably would not have been successful.

SO WHAT DO I DO ALL DAY?

After landing the education job, I had a whole new set of challenges and skills to learn. I quickly learned that my job basically has two components—daily administration and long-term projects—that require different ways of thinking and different rates of energy expenditure.

Administration is a high-wattage activity, requiring lots of motion and human interaction. In contrast, projects get done at more of a slow burn, with much intellectual energy expended in the process of developing ideas and making plans. Working on projects is a lot like research in terms of the creativity and thought involved, while the challenge in administration is to work with our Division's management team to motivate staff and to coordinate the big picture of activities in our Division. My biggest challenge in this job has been learning to make the transitions between administration and projects efficiently so that my productivity does not crash.

Outreach is a major component of my job. It is very important that our Division be visible and accessible to our constituents. This means lots of time on the phone and e-mail, and it also means much travel around the country to visit schools and attend conferences. During the academic year, I probably average about one 4-day trip per month. On many of these trips, I give talks, and each trip requires varying levels of summary reports at the end. Travel, including preparation and wrap-up, is a time-consuming but critical component of the job.

My job is professionally rewarding for many reasons. I get credit and recognition for the work I do, both within AIP in terms of advancement, and within the physics community in terms of professional visibility from talks and publications. Professional recognition is particularly important because, although I no longer make scientific contributions to physics, I make contributions in advancing physics and physicists as a professional community.

As noted earlier, I spend considerable time on outreach, either writing articles or traveling to give talks. My recent activity has been focused on assessing the relationship between physics and society with particular attention paid to the undergraduate degree and its perceived lack of usefulness to students. Specifically, I am concerned that the 38-year low in the number of undergraduate degrees in physics is an indicator that students and employers no longer view physics as a useful course of study.

In my travels and writing, I attempt to persuade students and the physics faculty that physics (or science in general, for that matter) is the best training for a professional life in our technical society, and that as a professional community we should aggressively promote this notion to students and the public.

COMPARING THE WORK

The work style in my job is intense but steady for 8 to 9 hours per day. This, coupled with my ability to make daily progress on important projects, is satisfying and suitable to my life-style, because I am able to advance professionally and I still have time for life outside of work. My hours and my ability to make progress are also different in comparison to my previous jobs.

Research could be managed in a normal day, but there was always a subtext of pressure stemming from possible competition from another lab, and also the pressure to make progress in the name of next year's funding proposal. The upside of research, and a similarity to my current job, was that the ideas and outcomes had my name on them, making my performance measurable and recognizable as my output.

In Congress I worked hard on important issues, and from a Congressional perspective, I made significant contributions. For example, in the subcommittee for which I worked, we labored for about 9 months on a bill to reauthorize high energy and nuclear physics in the aftermath of Congress's killing of the Superconducting Super Collider. Our bill was eventually passed by the House of Representatives. As with all House-passed bills, this one moved over to the Senate for approval, but due to timing it died a slow death by neglect.

This is a typical story in Congress: people work very hard to craft legislation, yet relatively few bills are signed into law. The legislative process is slow and incremental, and while this may be part of the genius of our system of government, it can be frustrating for the individuals who want to see their hard work pay off maximally.

On the other hand, even though our bill never became law, the policy it laid out did have lasting effect. The death of the SSC forced high-energy physicists to work with Congress and the Executive Branch in setting priorities for the future of their field. This cooperation and the way in which the scientists formulated their recommendations resulted in a generous bump in funding that helped keep high-energy physics healthy during the post-SSC transition. In this instance, scientists working with Congress and the federal agencies crafted a rational national science policy.

Another frustrating factor about working in Congress is that most creative output, such as speeches and op-ed pieces, is perceived as the work of the Member of Congress, not the staff person who actually crafted the words. At AIP, meaningful progress on issues that affect real people can be made daily and I get credit for my work.

What's Next?

Few career trajectories are deterministic. As a college student in my first physics class, I had no vision of myself sitting in College Park 12 years later worrying about the state of the physics education. I do not think that an undergraduate can plot a strategy for becoming assistant manager of the education division of an organization. The process is too nonlinear and dependent on timing and the relationships between individuals to have an outcome that can be predetermined.

My career strategy in college was to proceed in a manner that kept my options as open as possible. After my B.S., getting a Ph.D. seemed like the natural next step in keeping with this strategy. In retrospect, this was a wise decision because of the intense professionalization that occurs in the Ph.D. process.

Similarly, the unpredictability of career planning also applies to questions of where I go from here. For a scientist with experience in policy and management, there are many organizations in the Washington, D.C., area that could provide good opportunities. Within any one organization, such as AIP, the opportunities for advancement exist but occur less frequently.

For those desiring a career in nonprofit management in a science-related organization, I would suggest that being a member of the represented community is key; hence the need for a Ph.D. However, while probably necessary, a Ph.D. is not sufficient. Growth and experience in nonscience areas, such as management or policy, are important for developing the worldview, maturity, and temperament required of the position.

What is the Ph.D. good for? The Ph.D. is good training for a professional life. For me, it was my entrée into the culture of physics and the first step to my current job. This argument only works after the fact, however. Others at AIP have landed their jobs via a different route than I, but among the physicists, the post-B.S. study is our common bond.

20

· · · · · · · · ·

SCIENCE AND PUBLIC POLICY:

Translating between Two Worlds

· · · · · · · · · · · · · · · · ·

David Applegate, Ph.D.

American Geological Institute

Science threads its way throughout the entire federal government. The current session of Congress has featured debates and legislation involving standards for ozone and fine particulate matter in the air, acceptable radiation exposure levels for nuclear waste disposal sites, steps to limit the impact of global climate change, and many other issues with a hefty scientific or technical component. In addition, Congress set funding levels for the federal agencies that support or conduct scientific research. The range of issues addressed by the executive branch and by government at state and local levels is similarly dependent on science and technology, yet very few of the policymakers or policy-level staff in any of these settings have a background in science and engineering.

Historically, policymakers have sought the advice of the scientific community on issues large and small, and there is a long history of scientists providing expert opinion. But there is an equally long history of frustration and misunderstanding between policymakers, who are looking for simple, straightforward answers, and scientists steeped in uncertainty, multiple working hypotheses, and emphasis on detail. Policymakers are used to making decisions based on available information, whereas scientists are loathe to make a final conclusion, knowing there is always more data to collect and analyze.

THE NEED FOR SCIENCE IN POLITICS

This communication gap, combined with the growing importance of science and technology in society as a whole, has created a need for scientists who can work at the interface between science and public policy. What began (and still continues) as an advisory role done "on the side" of a traditional scientific career has evolved into a career in itself. There is now a significant cadre of trained scientists who occupy the nebulous space between their colleagues in research and policymakers in Washington. Like scientists who pursue careers in the media, scientists in public policy are translators between two worlds, filling a critically important need.

The opportunities in science policy are diverse and are not easily defined. What constitutes a career in science and public policy (or put more simply, science policy)? A narrow definition would be those engaged in policy for science, essentially the management or administration of science itself. A broader definition includes all those working on science *in* policy, the input of scientific information into a wide range of policy issues and decisions that have a technical component.

Although understandably diverse, this group—call them science policy wonks—share a number of qualities: they are good communicators, particularly as writers, they have an interest in issues outside their discipline, and they work well with a variety of people. The analytical skills developed through scientific training are equally useful in addressing policy issues. One congressional fellowship program for scientists listed desirable attributes for candidates as being not only a broad scientific background and strong interest in applying scientific knowledge toward societal problems but also a high tolerance for ambiguity!

Science policy is about advocacy, analysis, and advising. Scientists pursuing careers in science and public policy write speeches for members of Congress, develop environmental initiatives at the White House, manage federal agencies, prepare long-term policy analyses at think tanks, advocate on behalf of their colleagues at scientific societies, provide issue briefs for advocacy groups, unravel regulatory requirements at consulting firms, provide technical expertise for law firms, and engage in a host of other tasks for other entities.

Because of recent budget cuts and the expectation of future ones, the scientific community is realizing that it needs to focus on Washington and work harder to justify its share of federal dollars. There is a growing recognition that an important niche exists for scientists interested in fostering communication between their community and the policymakers. Although professors still remain who see value only in cloning themselves, many more recognize that there is enlightened self-interest involved in putting scientists into policy positions.

MAKING THE SHIFT

Unlike their counterparts in academic or industrial research, scientists in public policy often occupy jobs that are not exclusively reserved for their skills. Discussions over an academic faculty position might center on whether to hire a biological oceanographer or a physical oceanographer—a choice between a scientist and a lawyer is unknown. But many science policy positions can benefit from either of these backgrounds. Consequently, the scientist must make a convincing case for why his or her particular background is not only relevant, but crucial to success in that position.

To use the analogy of a symphony, lawyers are the violins of public policy, since we are talking about formulating the nation's laws. Scientists and other technical specialists are the woodwinds, adding another dimension to the sound of the more populous strings. Although few in number, scientists involved in policy are not equivalent to Woody Allen's marching-band cellist in the movie *Take the Money and Run*. Far from being out of place, their expertise brings a critical and necessary element into the process.

In the past, science policy was something that a scientist came to late in a distinguished research career, either by virtue of getting "kicked upstairs" into administration or being asked to sit on advisory committees or other "blue-ribbon" panels that are still an important facet of the science policy landscape. These tasks were done as an aside, and were not a career in and of themselves. Just like George Washington's ideal of a citizen-legislator, the scientists were expected to return to their laboratory after dispensing the necessary wisdom.

Today, those considering a career in science policy do so at many stages in their careers, sometimes after only an initial training in science. For example, an undergraduate biology major who minored in government or interned for a home-state senator and subsequently lands a job on Capitol Hill is not uncommon. These people become part of the broad spectrum of professionals whose preparation includes a solid foundation in science, bringing a scientific orientation and some knowledge of how science works with them to their job.

For a growing number of scientists, a career in science policy is undertaken following the many years of training and research leading to a doctorate. Some are drawn by a desire to see their training put to more immediate use, and others by outside interests in political issues (such as environment laws, health care, or education). This chapter focuses on the postdoctoral route because it is the path I took and hence the one most familiar to me. But I also will attempt to address how to get started for both those at the undergraduate and graduate (or postgraduate) level as well as future opportunities and challenges for a science policy career.

MY OWN PATH: TURNING A CONGRESSIONAL FELLOWSHIP INTO A CAREER

My route into science policy came through one of the few fixed points of entry: the Congressional Science and Engineering Fellowship Program. This program is essentially a 1-year postdoctoral appointment spent on Capitol Hill working with members of Congress or congressional committees as special assistants in legislative and policy areas requiring scientific and technical input.

Although administered by the American Association for the Advancement of Science (AAAS), the bulk of the congressional fellowships are funded by other scientific and engineering societies in a wide variety of disciplines. My fellowship was funded by the American Geophysical Union. Most require a doctorate or a master's degree with several years of subsequent experience in science or engineering. The sponsoring society is responsible for selecting the fellow and paying that person a stipend (which varies between $35,000 and $45,000). With their salary paid, the fellows come free-of-charge to work in the personal office of a senator or representative or for a committee.

This fellowship program was launched in 1973 in response to the lack of scientists on Capitol Hill and the perceived need for increased technical input in the legislative process. Between 25 and 30 fellows are chosen each year from a variety of disciplines. The program includes an intensive orientation on congressional and executive branch operations and a year-long seminar program on issues involving science and public policy.

The fellowships have launched many careers in science and public policy. Alumni of the program are well-placed throughout the Hill as staff directors, analysts, and professional staff members, as well as in high-level executive branch positions. After their fellowship year, about one-third of the Fellows stay on the Hill or elsewhere in the Washington area in some policy-related activity. About one-third return to their academic or industrial origins, and the remaining one-third make some other kind of career change following their fellowship.

I applied for the fellowship while in the final stages of completing my doctoral work in geology, studying the tectonic processes that shaped the

Death Valley region of California. My field area was about 30 miles from Yucca Mountain in Nevada, which the Department of Energy was studying as the likely site for the permanent disposal of the nation's high-level nuclear waste. I was intrigued by the technical aspects of the siting process—scientists trying to determine the potential for future earthquakes, volcanoes, or shifts in the water table that could create exposure pathways for the buried waste. My interest also was piqued by the difficult relationship between the scientists, the Department of Energy, and the many advocates for and against the site's suitability.

In addition, I have always been the sort of person who reads the paper every morning and tries to stay abreast of political issues. Having been a late convert to geology in college following several years as a history major, I still maintained a strong interest in the body politic as well as the planet Earth. The congressional fellowship was an opportunity to unite my diverse interests.

Plus, it was a job. When I obtained the fellowship, I was in a short-term postdoctoral appointment in my advisor's laboratory. Although I had a prospect for another postdoc out in California, it was not yet certain and I had already experienced a number of rejected applications for faculty positions that went instead to those who had already put in several years as an itinerant postdoc.

As a job, the fellowship could not have been better. Following an orientation into the ways of Congress and the federal agencies, I interviewed with a number of offices, meeting senators and representatives along the way. With my stipend paid for by the sponsoring society, the job market was considerably loosened!

I ended up with the Senate Committee on Energy and Natural Resources working on issues including Yucca Mountain, environmental cleanup of the nuclear weapons complex, the fate of science agencies in the Department of the Interior, dismantling the federal helium program (you probably didn't know we had one), and revision of mining law. With the partial exception of the Yucca Mountain issue, none were within my scientific expertise but all required me to be able to process scientific information and then provide it in a usable form for the senators as well as the committee staff.

As my fellowship year drew to a close, I decided that I wanted to stay in Washington and find a job in science policy. As luck would have it, my current job opened up, and my fellowship experience was critical to landing it. Following a brief stint as a regular staffer for the committee, mostly tying up loose ends, I became the rock lobbyist, an advocate for the geosciences with the more formal title of Director of Government Affairs at the American Geological Institute, a federation of geoscience societies. The job is a two-way street, seeking to inform and influence policymakers on issues affecting the geosciences and at the same time informing the scientists about policy issues in Washington that affect them. I do not own a single pair of Gucci tassel loafers, but I am a registered lobbyist.

AGI's Government Affairs Program is relatively young, founded in 1991 and still a small shop—two full-time and one part-time staff. My job includes a number of components that would be classified as lobbying—providing testimony before Congress and meeting with congressional staff and agency officials, both seeking and providing information. I also coordinate workshops that bring scientists to Washington to meet with their representatives.

A lot of time is spent writing—writing official letters, testimony, a monthly magazine column, electronic mail updates, and summaries of various issues. I am responsible for maintaining a web site with information on environmental, natural hazards, resources, appropriations, and other policy issues that impact the geosciences. I also try to get outside of Washington, meeting with member societies and giving colloquia at university geoscience departments.

GETTING STARTED: WHAT SKILLS DO YOU NEED AND WHERE DO YOU GET THEM?

Jobs in science policy require many of the same skills that make one a good scientist—research and analytical skills, the ability to handle multiple tasks, the ability to clearly communicate the results of your work, and a genuine interest in the subject matter. On top of that, you need the ability to work in a constantly changing environment. In most cases, you do not control the issues that come up.

A high premium is placed on the ability to write and speak well to a general audience and to work to a deadline. "People" skills are critical along with an ability to work with a diverse set of viewpoints. My job also requires management skills in order to maximize the productivity not only of myself but of those working with me. I still use many of the computer skills acquired through osmosis during graduate school, if not the computer languages themselves. Those skills are useful for coordinating our web site, navigating contracts with Internet service providers, and coding and designing web pages.

Where do you acquire these skills? Many come directly from one's scientific training, although the ability to write clearly and persuasively about technical issues for a general audience is sadly not a common component of scientific training at the undergraduate or graduate level.

Other skills may be acquired from unexpected sources or avocations. Some of my most valuable experience came from working on and eventually running a newspaper in college. In contrast to the semester-long gestation periods for term papers or the even longer timescale for research projects, the experience of having to get a finished product out the door each week taught me about writing under pressure, editing and fact-checking, working as part of a group, managing scarce time, and a host of other skills that upon reflection have served me well in later life.

Practice writing for a general audience in a school newspaper or write letters to the editor of the local paper. Learn to work to a deadline, and learn to work hard. One skill, however, can only be acquired once you have arrived in Washington, and that is a high tolerance for acronyms (to match the high tolerance for ambiguity mentioned earlier). Maybe military training would help.

For the job applicant, it is important to realize that policy jobs are outside the research/academic sphere. A curriculum vitae is handy to have in reserve, but the resume is the active tool for job-seeking. There are a number of excellent discussions about building resumes for nonacademic jobs such as Peter Fiske's *To Boldly Go* (Fiske, 1995). As suggested above, one's nonscientific experiences may prove as useful as accomplishments in the lab, and the resume should reflect expertise in computers, writing, and other skills with broad applicability.

Apart from specific skills and resume techniques, the scientist seeking a career in science policy must approach this new field with humility and the sense that they are there as much to learn as they are to teach. As much as you have to offer in technical expertise, the policymakers and lawyers with whom you will work can offer in policy experience and political savvy. Some ways to develop skills and experience follow, and sources of additional information can be found at the end of this chapter.

Internships

For those at the bachelor's or master's level, there are many summer internship opportunities in Washington (both paid and unpaid) to work for Congress, federal agencies, nonprofit organizations, think tanks, interest groups, and other Washington entities. In the case of congressional ones, they will not likely involve science-related issues but will provide an education in how Congress and politics in general work. A number of scientific societies, including my own organization, offer summer and semester-long internships in science and public policy. Some of these internships are paid and can represent a summer job. Many colleges also grant course credits for internships. State and local governments also have intern opportunities closer to home for many people.

Volunteering

Volunteering for a political campaign is another very different way to establish a network of contacts—when they need a scientist, they will think of you. A fellow grad student at MIT became involved in Democratic politics in Massachusetts, worked for the Clinton/Gore campaign in 1992, and ended up as a senior appointee to the AmeriCorps national service program and several other posts. For those who do not feel comfortable with politics in its undiluted form, there are opportunities to volunteer with advocacy groups and science organizations that have more of an issue focus.

More School!

Because most science policy jobs require a master's degree or higher, acquiring the necessary skills may mean going back to school for an advanced degree in science or in science policy. For those already in an advanced degree program in science, consider adding a component to your thesis that addresses policy implications for your research. For example, a graduate student in hydrology looking at groundwater transport could add a chapter on the use of groundwater models in assessing contamination potential at Superfund sites.

A number of universities now have programs in science policy, particularly environmental policy. A master's degree in policy combined with a strong scientific background could make for a very effective and relatively unique combination of skills. Many of these programs are geared toward working professionals, and it is possible to complete them while holding down a regular job. Although I do not want to advocate further proliferation of the legal profession, a science background combined with a law degree is another highly marketable combination in the policy world and elsewhere.

Whether in a policy program or in a straight science program, students with an interest in science policy should maintain a general awareness of what goes on in the world. When I was thinking about taking the Foreign Service exam, the advice I received from a long-time diplomat was to read *The New York Times* every day for several years before taking the test, and I would do fine. Similar advice applies here. A general familiarity with the mechanics of government and the issues of the day will limit the amount of nontechnical information on which you will need to get up to speed and will minimize the likelihood of appearing to be a cello in a marching band.

Fellowships

The congressional fellowship program described above is an important source of experience for scientists at a postdoctoral or midcareer level, but there are many more fellowship programs sponsored by AAAS and other organizations. Like their congressional counterparts, these fellowships typically are directed at those with graduate degrees in science or engineering, they last for one year, and they provide scientists with the public policy experience necessary to obtain more permanent jobs. At the same time, they are an opportunity to make practical contributions to the more effective use of scientific and technical knowledge in the U.S. government, and to demonstrate the value of science and technology in solving important societal problems.

Other AAAS-sponsored fellowships include the Science, Engineering, and Diplomacy Fellows Program in which fellows work in international affairs on scientific and technical subjects for one year, either at the U.S. State Department (1 fellowship) or U.S. Agency for International Development (12 fellowships). A Risk Assessment Science and Engineering Fellows Program places scientists at the Environmental Protection Agency (5 fellowships) and U.S. Department of Agriculture (1 fellowship) for a year as special research consultants, assessing the significance of environmental and/or human health problems through application of risk assessment.

Other fellowships include one at the RAND Critical Technologies Institute, providing expertise in research and development, technology transfer, international competitiveness, and related issues. The newest AAAS-sponsored program places scientists in the Department of Defense (5 or more fellowships) for 1 year to work on issues related to defense policy, technology applications, defense systems analysis, and program oversight and management.

Some scientific societies, for example the American Physical Society, offer science policy fellowships that bring young scientists to work with the society's own government affairs program, again intended to provide the necessary policy expertise to complement the person's considerable scientific expertise.

A number of other fellowship opportunities are not specifically for scientists, but may be open to them. Two similar programs on Capitol Hill are the Robert Wood Johnson fellowships for medical doctors and the American Political Science Association fellowships for (you guessed it!) political scientists. Scientists who are employed at federal agencies can participate in the LEGIS/Brookings Institution Fellowship program, which places them in a congressional office for up to 2 years.

The White House Fellowship is another program to consider. Started in the Johnson Administration, the White House Fellowship recruits outstanding up-and-coming professionals from all disciplines to serve for 1 year as a special assistant to the president, the vice president, or one of the Cabinet secretaries. Among more than 500 alumni of the program, only a handful are scientists. Scientists have had a record of outstanding success in the program, but few apply, probably because it is poorly advertised in the scientific community.

The Presidential Management Internship (PMI) is another outstanding opportunity for people right out of school. Every year, PMI takes from 200 to 400 people into a 2-year program that places them in Cabinet agencies, the White House, and Congress on rotational and developmental assignments. The goal of the program is to develop the government administrators and leaders of tomorrow.

At the end of the 2-year program, PMIs automatically are converted to a permanent position in their agency, a unique aspect of the program. Scientists have been very poorly represented in this program as well, but the Office of Personnel Management and particularly the Department of Defense are eager for more applicants from science and engineering schools.

Most people apply right after a masters degree, but some come in after doctorates as well. The starting salary is on the low side ($31,000), but PMIs receive regular raises during their 2 years and finish the program at around $45,000. Often they move into a higher position after their internship.

POINTS OF ENTRY INTO A SCIENCE POLICY CAREER

Given the diversity of opportunities relating to science policy, there are consequently many trailheads that lead into science policy career pathways. Those at the earliest stage of their training as scientists have the greatest variety of paths to follow, because they have more opportunities to adapt their education and are more flexible about salary requirements (i.e., will work for food).

For those who already in the midst of their careers, an entirely different set of opportunities exists for developing the necessary skills and experience. You can become known among decisionmakers by demonstrating a willingness to communicate your expertise through your participation in National Research Council studies or on advisory boards for federal agencies. Certainly, the administration of science agencies is an opportunity for those who seek it, just as a deanship is an opportunity for those professors with an interest or flair for administration (or a knack for herding cats!).

This section reports on a number of possible pathways for people with a science background and experience gained through the internships, coursework, fellowships, and volunteering. All of these points of entry together represent only a handful of the many ways into science policy careers. The following paragraphs cover nongovernmental organizations (NGOs), Congress, and federal agencies in descending order of opportunities and experience levels required.

Washington is teeming with a motley assortment of nongovernmental organizations that occupy a somewhat nebulous zone between the public and private sector. Naturally, they are a haven for science policy wonks. Most of these organizations hold nonprofit status, although some may represent for-profit entities. NGOs tend to have fairly flat organizational structures. Although limiting the potential for advancement, these structures provide entry-level staff with a host of opportunities and responsibilities that more hierarchical structures (such as the Department of Energy, where it is possible to hold a job 20 layers beneath the Secretary!) do not allow. Several categories of NGOs have opportunities in science policy, including scientific societies, think tanks, and interest groups, and the National Research Council.

Scientific Societies

Many scientific, engineering, and professional societies as well as university consortia are based in Washington or have Washington-based government affairs programs. Many of the jobs are filled by individuals without a background in the scientific discipline they are representing. My own job falls into this category and is described in detail above.

The size of these programs varies from 1-person shows to as many as 20 people. Scientists who apply for these jobs generally have some policy experience either in Congress or with a federal agency, part of the "revolving-door" world of Washington.

As with other types of science policy positions, these jobs are not reserved for scientists, and a scientist applying for such a position will be competing against other candidates who have considerable experience in government affairs as a career in itself, often having advocated for a variety of different issues and organizations.

I would argue that ideally scientists should be representing the scientific community. Otherwise, the advocate becomes another buffer between the scientists and policymakers. Having a background in the discipline that one is representing also provides insight into how issues concern or impact scientists, and may improve other scientists' comfort level with the advocacy process.

Think Tanks

Another type of NGO is the so-called "think tank," places like the Brookings Institution or Resources for the Future that provide policy analysis and advice to government and industry. Although these organizations tend to be dominated by social scientists, they address a range of technical issues that require a strong scientific background.

While most of these organizations view themselves as nonpartisan, others (such as the Cato Institute) occupy a particular region of the political spectrum. A number of scientific societies also have policy-related programs that fall into this think tank category. Rather than advocacy, such programs focus on increasing the use (and usefulness) of scientific information in the policy-making process. An example is the Ecological Society of America's Sustainable Biosphere Initiative.

Interest Groups

That leads us to yet another group of NGOs, known as interest or advocacy groups. One need think only of the Sierra Club or the National Rifle Association. Trade associations are a related group, often employing policy analysts and others to help make their case. Some groups such as the Environmental Defense Fund pride themselves on the number of scientists that work for them.

Although some of these jobs are in research, most are policy related or have a significant policy component. Working for a group with a clearly defined agenda may make some scientists uncomfortable, but for someone with a strong affinity for a given issue, working for an interest group can be very rewarding. For better or for worse, interest groups play a dominant role in informing and framing public policy debates and thus represent a tremendous opportunity for scientists who wish to improve the technical basis upon which these groups advance their positions.

National Research Council

The National Research Council is the principal operating arm of both the National Academy of Sciences and the National Academy of Engineering. Although it is congressionally chartered, the Research Council is not part of the federal government but it does derive the bulk of its funding from Congress and from federal agencies. Much of the Research Council's activity involves the preparation of reports by committees consisting of National Academy members and other experts on a variety of science policy issues.

Although committee members are volunteers—one of the principal means by which scientists provide advice to the federal government—the Research Council employs a considerable number of doctoral scientists to staff the committee studies and to develop new ones. These staff officers have backgrounds across the entire breadth of science and engineering disciplines. The staff plays a major role in shaping the reports and consequently in contributing to significant societal decisions.

The qualities that are most important in this post include a willingness to work in fields outside your scientific discipline, a love of learning about new fields, the ability to work well in teams, and, above all, being a good communicator who enjoys and excels at writing. The positions represent an opportunity to help define the agenda and to interact with all-star casts (hence a certain deference may also be a useful skill). The Institute of Medicine is the equivalent agency for the medical community with similar opportunities for those in health-related fields.

Congress and Congressional Agencies

Like NGOs, Congress has a very flat organizational structure. In a congressional office, even the interns are no more than three levels below the representative or senator for whom they are working. As a result, staff have a tremendous amount of responsibility and potential impact on public policy. Moreover, the turnover rate is high—the average tenure of a staffer working in a House personal office is under 2 years—so advancement can be rapid. The average personal office staffer is in his or her twenties, holding a bachelor's degree. Many worked on campaigns or interned before obtaining a staff position.

As an example, one of my program's interns had just graduated from college with a geology degree. Based on her internship experience and an earlier internship on the Hill, she obtained a job with her home-state senator as a legislative correspondent, answering constituent mail. Several months later, her boss became chairman of a committee, and she landed a job working on science education issues for the committee.

Committee staffers generally are older and experience lower turnover. They also are more likely to have a law degree or other advanced degree but few have a technical background.

A number of scientists work for the House Science Committee, House Resources Committee, Senate Energy and Natural Resources Committee, Senate Labor and Human Resources Committee (science education and biomedical research), and Senate Commerce, Science, and Transportation Committee. Although many are former congressional fellows, others arrived from federal agencies or came to work in a personal office and moved up with time. In turn, many congressional staffers obtain policy-level jobs in federal agencies based on their Hill experience.

Opportunities also exist in two congressional agencies—the Congressional Research Service in the Library of Congress and the General Accounting Office. The former is the principal source of information for Congress on every issue imaginable, including scientific issues and related policy areas such as the environment, natural resources, and health care. The latter is a watchdog agency, undertaking audits and other investigations into activities at federal agencies and elsewhere at the request of Congress. Although dominated by social scientists and assorted number-crunchers, the GAO addresses a wide range of issues that require a technical background.

Federal Agencies

In the United States, science is dispersed across many mission agencies with research conducted in support of program goals. That dispersal means that there is potentially a much broader set of opportunities for science policy wonks. In the European model, a single centralized science agency provides a single entity for research funding, giving scientists more influence in their own affairs but not beyond.

Although federal agencies are a natural home for science policy wonks, there are few points of entry. Instead, most scientists who have agency policy jobs either came up through the ranks from technical positions or came from jobs in Congress or NGOs. There is a variety of positions in science agencies and mission agencies in legislative affairs or as special assistants to high-level political appointees. Consequently, many of these policy-level jobs are themselves political appointments.

Although officially the hub of science policy in the federal government, the White House Office of Science and Technology Policy has relatively few opportunities for direct employment. Instead, scientists either arrive at OSTP with outside funding (such as a fellowship), on loan from their host agency, or as a political appointee. Scientists at OSTP are there either because they are experts on a specific hot issue or because they bring a broad background and understanding. In both cases, they have to be capable of writing speeches that contain good sound bites but keep the science accurate.

WHERE DOES A CAREER IN SCIENCE POLICY LEAD?

It may seem like an odd selling point for a career pathway, but a benefit of science policy is that it can lead all sorts of places, many of them not actually *in* science policy. As indicated earlier, many who do congressional fellowships take their experience back to their earlier jobs or go in entirely new directions.

Just as academic training is the coin of the realm for the newly minted scientist, so policy experience is valuable, whether going back to academia, industry, or into a nonpolicy government research position. There are also opportunities in state governments at agencies handling environmental, health, resources, or technology issues. Having a firm understanding of the forces that are at play in obtaining grants, getting attention at the agency level or at the White House, and knowing how the political winds are changing are all invaluable to a business, government laboratory, or university department that relies on federal support.

For those who would like to return to academia (if not necessarily as a researcher), it is possible to teach a course or two as an adjunct professor in one's scientific specialty or on science policy. There are also a growing number of programs in science policy, and it may be possible to secure a full-time position in the field, returning to academia in a different field from the one you were trained in.

For those who get the Washington bug and want to stay in science policy, a number of the points of entry discussed above can lead to long-term appointments. But these seldom lead to a lifetime job. As in most other fields, mobility and multiple careers is the norm for scientists who become science policy wonks. The high turnover rate in Congress provides opportunity but not necessarily continuity. Political appointments in particular have a way of vanishing in the wake of elections.

Although the flat structure of many NGOs may not be particularly conducive to upward mobility, opportunities do exist in management. Many of the same qualities that make for a good advocate are also valuable for leadership positions. Furthermore, experience in government affairs at a scientific society or interest group is readily portable and can lead to similar work for the private sector, federal agencies, or other organizations.

Law firms increasingly represent another potential avenue for scientists with policy experience. Just as Congress deals with highly technical issues in the process of making laws, the same holds true for those who interpret and argue over those laws. Many firms hire economists, engineers, and scientists to assist with the technical aspects of cases in areas such as environmental or intellectual property law.

Many of the opportunities that have been described focus on scientists providing advice and analysis to policymakers, but there is no reason that a science policy career should not lead scientists to become policymakers themselves. In contrast to the droves of lawyers and businesspeople in Congress, very few senators or representatives have a scientific background.

Their small numbers include one geologist, one physicist, one chemist, and a couple of engineers, doctors, and veterinarians. In the past election, two former science fellows ran for Congress. Although neither won, there are many races yet to run.

EARNING POTENTIAL

The senior advisor for my program likes to say that our organization is not-for-profit and good at it! For the most part, science policy types labor firmly in the middle class with pay scales similar to those of their research counterparts in academia and government. That said, however, the earning potential can be significantly greater for those in the upper echelons of federal jobs, including senior executive service positions and political appointments, which can pay in six figures. For those who leave the nonprofit or government sectors and work for government affairs offices in industry, compensation can likewise be considerable.

CONCLUSIONS

At a time when the emphasis among career planners is to prepare for not just one career but many, science policy is an important and fascinating choice to consider either as a diversion from one's present course or as an end in itself. Because it is broadly defined, many of the opportunities are there to be made. Science policy is a career field consisting of many niches, some of which are not readily apparent because the case for them must be made to other scientists and to policymakers alike. Although the lack of ready-cut positions and lines of advancement may seem daunting, any career field where you can make your own niche is one with a lot of growth potential and freedom to tailor a position.

Like one's education, policy experience is money in the career bank. The combination of scientific expertise and a firm understanding of how government works is valuable and marketable for jobs at consulting firms, large companies, universities, and within federal or state agencies.

Scientists are capable of much more than the specific laboratory techniques that they perfected in graduate school, and there is a very real need for scientists to apply their skills and knowledge to the public policy-making process. In doing so, they not only may find a satisfying alternative career for themselves but will help to ensure that traditional science careers for their peers do not vanish under budget pressures for lack of a compelling justification.

ADDITIONAL SOURCES OF INFORMATION

On-Line

Additional information on fellowships is available from the American Association for the Advancement of Science (http://www.aaas.org), the American Political Science Association (http://www.apsanet.org), and the Robert Wood Johnson Foundation (http://www.rwjf.org).

Contact information for Members of Congress (http://www.house.gov *or* http://www.senate.gov) or the White House (http://www.whitehouse.gov).

Information on staff and internship opportunities at the National Research Council and Institute of Medicine (http://www.nas.edu).

A good on-line listing of graduate programs, including science and public policy, is available from Peterson's (http://www.petersons.com).

For one example of the type of policy analysis and advocacy done by government affairs programs, visit the American Geological Institute (http://www.agiweb.org).

For more information on think tank activities, visit The Brookings Institution (http://www.brookings.org) or Resources for the Future (http://www.rff.org).

Hardcopy

Fainberg, Anthony, ed. 1994. *From the Lab to the Hill: Essays Celebrating Twenty Years of Congressional Science and Engineering Fellows.* Washington DC: American Association for the Advancement of Science.

Fiske, Peter S. 1996. *To Boldly Go: A Practical Career Guide for Scientists.* Washington DC: American Geophysical Union.

Smith, Bruce L. R. 1992. *The Advisors: Scientists in the Policy Process.* Washington DC: The Brookings Institution.

Stine, Jeffrey K. 1994. *Twenty Years of Science in the Public Interest: A History of the Congressional Science and Engineering Fellowship Program.* Washington DC: American Association for the Advancement of Science.

Tobias, Sheila, Daryl E. Chubin, and Kevin Aylesworth. 1995. *Rethinking Science as a Career: Perceptions and Realities in the Physical Sciences.* Washington DC: Research Corporation.

21

• • • • • • • • • •

RESEARCH FUNDING
ADMINISTRATION:

Matching Money with Research

• • • • • • • • • • • • • • • • •

M.J. Finley Austin, Ph.D.

Administrative Director, Merck Genome Research Institute

My decision to leave the lab and academia was not easy for me. Although I long held an interest in policy and administration (I am a big picture, visionary type of person) and saw myself in administration some day, my departure came more out of necessity than immediate desire at the time. I would like to add, however, that I have no regrets. In fact, at this point I am glad life kicked me in this direction when it did.

I always loved school and learning and science. As a young child, I had already set my sights on getting a higher education despite the fact that this was not considered a high priority in my family. Supporting myself, I went directly to college in 1976 after graduating from high school. My goal was to go on to medical school. At the time, medicine was the only job I was familiar with that was scientific in nature. When I got to college, I soon realized that there was this unbelievable, exciting research enterprise in universities and my goals began to change.

During my sophomore year I changed my major from biology (premed) to psychology (not viewed as very practical by many, but it satisfied a long-standing interest in human behavior) and I graduated with a B.S. in 1980. Still uncertain as to what I wanted to do next, I went to work full-time at the emergency psychiatric clinic, where I worked part-time during my senior year. I also continued to take classes, some graduate psych as well as the necessary requirements to reopen the option of applying to medical school.

I'll spare the reader the details, but I continued to vacillate for several years about what to do next. I wanted to continue my education, but I remained uncertain as to the direction. I changed jobs in 1982 and started working as a lab tech in a neuropharmacology lab. I have always had a strong desire to do something to help people, but in this new setting I soon found I was much happier doing research with rats than working directly with patients. This got me thinking more about graduate school as an option, rather than medical school.

After a few years and several fits and starts with filling out graduate applications (most only read and never completed, much less mailed) for various programs, and some real-life adventures, I became serious about beginning graduate school. Thanks to two books, *On Human Nature* and *The Genetic Prophesy*, I decided human genetics was the field for me. In 1984 I applied for and entered a Ph.D. program in human genetics. I was especially glad I had taken all those biology and chemistry courses after receiving my B.S.; otherwise, it would have taken me at least a year of course work before I could have even applied.

I had ups and downs in graduate school, but I always loved my research. After working in a university research environment, I was convinced that there was no better life than being an academic researcher and professor. This became my singular goal. I defended my dissertation and went straight into a postdoc, my sights still set on the road to becoming a faculty member at a research university.

I was, however, not completely and blindly driven by doing research. It was not my life 24 hours a day, 7 days a week (more like 12 hours a day, 6 or 7 days a week). Personal concerns played a role in my career decisions. Therefore, I wasn't always able to play by the normative rules of how to get ahead in academic science. Both of my degrees come from the same university and I stayed there for my postdoc as well, though in a different area and department. These choices, like all choices, have impacted my career, but not nearly to the negative degree conventional wisdom would lead you to believe.

My postdoc was going well, despite my suffering a great personal loss during the first year. I applied for and was awarded a fellowship from the National Institutes of Health. My work was progressing and I started job hunting in the third year of my fellowship.

Even though two postdocs seemed to be becoming the norm in the biomedical sciences, I aimed for my dream—a faculty job. I knew a position at a top institution might be unlikely, but I sincerely did not imagine I would have any problems landing a job in an academic research setting at a credible state school. It was now 1991, the alleged beginning of a Ph.D. shortage that had been predicted in the mid-1980s by the National Science Foundation (NSF), which should have translated into lots of unfilled faculty slots.

Unfortunately, the NSF's predictions didn't pan out for me. About half the universities to which I applied never bothered to respond to my applications, and those that did respond would state without fail that the position had attracted hundreds of applicants and that I should not take my rejection as a reflection of the quality of my work. I applied to get my fellowship extended, taking me into the fourth year of my postdoc.

In year four, the job hunting did not improve even though I started broadening my efforts to include industry research jobs and faculty positions at liberal arts colleges. I got a few nibbles for independent positions, but the only concrete possibilities were for second postdocs. I was becoming increasingly concerned about my personal future, as well as the future of the research enterprise. I just couldn't understand how the NSF could have been so far off the mark. I knew it wasn't just me: at more and more of the large national science conferences I attended I heard more and more postdocs, many with better academic pedigrees, telling the same bleak story.

This dearth of jobs, in the face of predictions by the NSF, led me to pay increasing attention to science policy issues. Much to my surprise, many leaders in the research community were not particularly interested in what was happening to younger scientists' career development. Quite frankly, I started getting angry about this and wanted to find a way to direct my energy and make a positive impact.

I Leave the Lab

Early in my postdoc, I had read in *Science* about science policy fellowships offered by the American Association for the Advancement of Science (AAAS). At the time, I put the article aside thinking it was something I would try to do on sabbatical after I got tenure and had a well-established research program. As my job hunt failed to produce, and the thought of a second research postdoc continued to sound like an undesirable idea, I decided to apply. My rationale—it looked really interesting, sounded like a lot of fun, and if I had to take another temporary fellowship I was better off doing something that would give me more options than would staying in the lab. Also, the AAAS fellowships pay a much better stipend than a typical academic biomedical research postdoc. This translated to a 66% increase in stipend if I got the fellowship, certainly a good selling point. I submitted applications to several programs—I was offered a fellowship through AAAS to work at the U.S. Agency for International Development. I accepted.

Several faculty members at my academic institution tried to discourage me from taking the AAAS fellowship and leaving the lab. When it became clear I was going to Washington, D.C., many told me that I could get away with leaving the lab for a year and still come back, but that I would never survive 2 years out and be able to return to research without great effort. I should add, however, that not all reactions were negative, that there were several faculty members who were very supportive and a few who were even envious. I was excited about going to D.C. but equally scared. In my heart, I knew I wouldn't come back to the lab after a year because the only way to do that would be to take another postdoc. I knew I wouldn't do that unless it was the only option open to me. I was simply getting too old to continue in temporary positions with minimal benefits.

Before I go on, I would like to say that I think doing research is great, exciting, fun and a vital experience for a student of science. If a new type of senior postdoc position had been created in the academic biomedical sciences at that time—one that paid reasonably, provided real benefits, and offered some tangible level of stability and independence—I might very well still be working in a lab somewhere.

But I do not regret leaving the lab; what I do now is just as exciting and fun! I am having a great time and wouldn't trade my job now for a faculty position (even if it paid as much). I should also point out that the extra research experience I gained as a postdoc—4 more years of research, competing for funding, and learning molecular genetics—is critical to doing my current job well.

So off I went to Washington with very mixed emotions. My first night in town, I cried as I sat with my dog on the floor of my empty apartment awaiting my furniture delivery the next day. I called my best friend from grad school and told her what a terrible mistake I had made—I had ruined my life, because I knew I'd never be able to return to the lab. I wondered how I could ever be happy now. It was more than the loss from giving up a dream or vision of my future—I love science, research, and academia, but I felt my love was unrequited.

Obviously, my life wasn't ruined. And, in one of life's great twists, my friend is now doing an AAAS science policy fellowship herself. It took me some time and I had a lot of feelings to resolve, but I began to see new possibilities in using my expertise to make a meaningful contribution to science. Also, I confess, I kept my hand in the lab for about a year. On many weekends I drove the 100 miles down to Richmond to finish my last project and get one more publication out. However, I really enjoyed Washington and, in time, the drive to Richmond got longer and the drive home to northern Virginia shorter. Any doubts I had about leaving the lab disappeared, but I did still miss academia.

Overall, my Washington experience is one of the best mistakes I have made. My AAAS fellowship work included grants program management, as well as other good solid practical administrative experiences. There were many opportunities to meet and work with a variety of people. Many of the vast array of people I worked with outside of the government came from research foundations with diverse missions such as funding U.S. research; helping third-world nations develop their science and technology infrastructures; and informing health care and science policy. My grant administration experience and exposure to these individuals made me realize I should be including this sector in my job hunt.

THE WORLD OF PROGRAM DIRECTORS

The first position I secured at the end of my 2 years as a AAAS fellow was with the Burroughs Wellcome Fund. The Fund is an independent private foundation that supports biomedical research. They advertised the opening in *Science*. Given my new experiences in science policy, the match sounded good so I applied.

The Fund hired me to serve as program officer for a number of competitive award programs providing funding in the areas of pharmacology, toxicology, and infectious diseases. My primary responsibility was the administration and oversight of the granting process (e.g., soliciting applications, planning advisory meetings, and interacting with grantees and university sponsored programs offices). Many times I helped plan special symposia at national meetings on topic areas that the Fund supported. I traveled extensively to conferences to stay abreast in current research areas and to highlight the Fund's activities. I enjoyed my work and had begun several special initiatives (e.g., a mentoring program to couple new investigators with senior leaders in their field) that were particularly fun and rewarding.

Furthermore, my job at the Fund offered unparalleled security (the benefit of a large endowment). However, I was willing to give this up after about a year and a half when I was offered my current position as administrative director of the Merck Genome Research Institute (MGRI).

That this opportunity opened up for me was truly an instance of everything coming together at just the right time. A general letter inviting nominations for the position was passed along to me. I thought I had a pretty good chance at the position, given my background in human and molecular genetics coupled with my nonprofit funding experience. I put a call into the president of the Institute. The more he told me about the position, the more I knew that I wanted this job. Although I was satisfied with my position at the Fund, I did miss being in the mainstream of genetics. Working for the MGRI lacked the long-term security of the Fund, but I could not pass it up—the Institute is wholly devoted to functional genomics research, the position was a step up, and it meant the chance to be at Merck!

You see, the pharmaceutical industry had become a place where I hoped to work someday and you can't get any better than Merck. By the time I left Washington, I had become active in the area of policy and science career development. I continued in these efforts after joining the Fund. For example, the AAAS invited me to serve on a special task force to study what the association could do to increase or enhance their career development undertakings, and I was elected to the board of directors of the Commission on Professionals in Science and Technology.

These activities brought me into contact with a number of individuals from the pharmaceutical industry. They helped change my negatively biased view about industry, acquired during my years in academia. For the most part, those I encountered from this industry were bright, caring scientists who liked their work. They took great satisfaction in the fact that the end results of their research really do make a positive impact in the world. I decided that at some point I would like to work for a major pharmaceutical company. Merck is not only a well-known leader in the industry, but I soon learned it has a long-standing reputation for being a great place to work.

I had begun talking with people in the industry to identify nonlab positions where my skills would be beneficial, but I never expected an opportunity to arise so quickly or to be so ideal. The MGRI is a not-for-profit institute established by Merck & Co. to support genomic sequence to function technology development.

Currently, a large quantity of genomic sequence is being generated for many organisms, including man. The MGRI goal is to support the development of basic research technologies to effectively and efficiently mine this rich information source. We support research in universities and other settings and we make the results of these efforts publicly and widely available. The Institute is separate from Merck & Co. and I play no role in the for-profit activities (I am a Merck employee on lease to the Institute), but I am in a position to be able to observe and learn about the pharmaceutical industry.

I still can't believe I was so lucky to have such an opportunity become available to me. I feel the same way that many of the faculty I know say they feel about research: I am having a great time and I get paid for doing it!

My actual situation in research funding administration is fairly unique. There are certainly a number of possible career avenues for research funding administration, but the establishment of this Institute by Merck is peerless. Merck has a long tradition of supporting the sharing of research results and tools to ensure that progress is not impeded downstream, which could prevent the development of new pharmaceuticals to benefit humanity. I know of no other pharmaceutical company that has established an independent foundation to fund public domain basic research at this level. The reader needs to keep in mind that my actual situation is rare, but comparable positions do exist in other settings (at the National Institutes of Health, NSF, other government agencies, and not-for-profits) and many of the work activities I describe apply in all of these environments.

Since the MGRI is housed at Merck & Co.'s facility in West Point, Pennsylvania, which is somewhat similar to an academic research setting, I am surrounded by bright, interesting people. The Institute itself is just a staff of two—myself and my secretary, but I am not isolated. Even though the work of the MGRI is completely separate from Merck & Co., I still have cause for interaction, both professionally and socially, with scientists and others in the company. Thus, I get the same kind of broad contact and intellectual stimulation as with an academic position.

THE JOB

In my position as administrative director I serve the Institute's Board of Trustees, all of whom are senior management within Merck from around the world, and I also work with our external advisors, a group of five renowned researchers from around the United States. I also interact regularly with the administrators and researchers at the institutions we support. I am privileged to work with a lot of talented, interesting people as well as many of the greatest scientific minds of our time.

I'll admit, at times there certainly are the frustrations that come with any job, and every large organization has its own bureaucracy to contend with. But I really have the best of both worlds: the ability to work rather independently on a large project without being isolated. Most of the work is collaborative in nature, so I have interaction with, and input and backstopping from, an array of bright individuals.

My day-to-day work environment is rather independent. Administering the Institute involves overseeing the solicitation of applications, reviewing them for appropriateness to the mission, monitoring the external and internal reviews, providing the board with all the necessary information to make final funding decisions, monitoring progress of awardees, and serving as project manager on several special initiatives. Other duties include budgeting, planning board and other meetings, establishing and maintaining a grant-tracking data base, and developing and maintaining a web site for the Institute.

I spend a lot of time on the phone, on the web, writing correspondence and other materials, and e-mailing—in other words, gathering and disseminating information through a range of methods. This includes providing guidance to potential applicants to ensure that we solicit only those applications that are appropriate to our mission; seeking information on current technologies and related funding efforts to help guide our program; arranging and attending the meetings that are necessary to run the Institute and its projects; and seeking potential funding partners when appropriate.

By necessity I must read broadly, including technical journals and the lay press, and I attend conferences. I need to stay abreast of the latest technological developments while working to highlight the mission and activities of the MGRI to attract appropriate, innovative proposals from outside scientists.

Travel opportunities (or requirements, depending on your perspective) are a necessity for funders, but they vary depending on where you work. There are certain conferences that you must attend, and site visits are extremely helpful for gaining a solid understanding of the work that is being supported. In my last position, my project portfolio included such a diversity of scientific fields that I had a huge number of meetings to attend and ended up traveling on average 2 to 3 times a month. Since the MGRI mission is more focused, there are fewer "must attend" activities and I get to spend significantly more time at home.

Many of the Institute activities mentioned above are backstopped and assisted by a number of Merck's divisions, predominantly legal and public affairs. On a weekly basis, I work most closely with two individuals other than the Institute secretary. One is the president of the Institute, Dr. C. Thomas Caskey, who also serves as my "Merck" boss and is a senior vice president within the company. The other is a Merck lawyer, Ms. Mary Bartkus, who serves as the MGRI's administrative secretary. Both are conscientious mentors.

SCIENCE AND FUNDING

My personal perspective is that funding science is like the scientific process/method itself—it is very dynamic. Research funding efforts are born of a long-term goal or vision, much like a theory. To accomplish this vision, a series of short-term projects (hypothesis testing) must be employed. Since it is science that is being funded, there is the constant need to assess the data and reevaluate to make certain that the short-term steps achieve the long-term goals. In order for me to understand and carry out my board's vision and the Institute's mission, I must understand the science. I have to be able to communicate information in both directions—out to applicants and back to the Institute. I believe I am a better administrator because of my scientific background. I bring critical, scientific thinking to planning and management, not merely scientific or technical facts to the job.

There are a number of reasons I enjoy and gain satisfaction from my job. One of the most rewarding aspects is being a part of creating the future of the program. It is a lot of fun starting something new. Plus, I like "producing" projects: figuring out the right questions that need to be asked to plan a project effectively and bringing the right people together to answer them and get the job done. I am a long-term thinker; I derive a great deal of satisfaction knowing that my efforts will have a sustained impact on science. Just as in academia, I get to think about topics and problems that I find fascinating, and to pose questions in an effort to understand them better.

For me, there are a few downsides of administration. There are occasions when I have to wear strict business attire. More problematic is that I do occasionally encounter the old attitude that I must somehow be a failure because I left academia. But even this does not bother me any more. After all, I am surrounded at work by people who left academia and are happy and thriving professionally. Furthermore, I know that my role is important and that I do a better job because I have first-hand experience in scientific achievements.

In essence, all the skills one needs to succeed in research are the same as those needed to succeed in funding administration—curiosity, self-motivation, constant learning, critical thinking, the ability to communicate concisely and effectively both orally and in writing, networking with others, and the ability to motivate others to work together (the funder, and not simply the money, can be the key to a successful collaborative project). Plus, it is important to develop a broader vision of the process that looks beyond singular research interests. There needs to be an appreciation of the big picture and an understanding of the many competing interests involved in the "enterprise" side of science.

GETTING THIS JOB

I find it a bit difficult as to how to advise someone to follow my path, since it is not what I originally set out to do—the slings and arrows of fortune certainly played a role in moving me through the maze. The path looks natural, but it really has been the sum of a variety of choices, nonchoices, and the acceptance of opportunities that were unimagined until their very availability was made known to me.

That said, I never pass up a chance to dispense advice. With the privilege of hindsight, my recommendations, above all, are to learn the science and to learn how to think critically. While in the lab, get your own funding whenever this is possible. Also, learn how to write things other than research papers (e.g., memos, meeting agendas, and general correspondence). Make the most of your training (e.g., write grants or mock grants, and use your committee meetings as a chance to really learn how to plan an agenda and conduct a meeting).

Do at least one postdoc—you'll need the added research experience if you want to be taken seriously by the researchers with whom you will work. But consider doing a postdoc in industry—it certainly pays better and typically you get full employee benefits. More importantly, it will broaden your perspective. Also, look into volunteering in the sponsored program and/or technology transfer office at your university or, better still, do part of your graduate assistant work in those settings. Nothing compares to having actual experience doing the job you are trying to get (conversely, you might find out this is not a road you want to travel before you commit a lot of time and energy).

Get involved in science policy at the state or local level. Working with a lobbying group will give you the chance to communicate science to a broad range of people. Finally, don't forget these activities when writing your resume, c.v., or cover letter.

Where Do I Go from Here?

Where do I see this job taking me? Right now I can honestly say, Hopefully, nowhere! I am quite content, but if I were to try to look into the future (an exercise I do not recommend) I see numerous possibilities. I am gaining experience that applies to a number of other areas here at Merck, such as industry and academic collaborations, and public affairs.

Outside of Merck, if given the appropriate opportunity for growth, I would be happy to continue my career in the private foundation world. I would gladly return to Washington, D.C., to work in a government-funding or policy-making setting. Working in a scientific society or organization is certainly a possibility. I could even see myself back in academia in a university-sponsored programs office or tech transfer office (although I much prefer giving money away to trying to get it), or administering graduate training and research. I have learned that there are a lot of scientists who could use some training in grant writing, so I could probably even return to teaching.

If you are reading this book, it is safe to assume you are looking for some new directions and ways to apply your scientific talents. I hope my story has been of some help to you during your decision-making process. It did not happen the way I imagined (which is why I now recommend against crystal ball gazing) and there were certainly some potholes in the road (and more to come, for sure), but I can say that all-in-all, science and education have served me very well. Apply the scientific approach universally and it will serve you well, too. After all, a good scientist always keeps an open mind to all the possibilities.

chapter

22

GOVERNMENT AGENCIES:

Directing Science in the Military

Genevieve Haddad, Ph.D.
Senior Program Manager, Directorate of Chemistry and Life Sciences
Air Force Office of Scientific Research, Washington, D.C.

I manage a $5 million basic research in biology program for the Air Force, which has a current focus on an area of neuroscience called chronobiology. This program supports basic research on the circadian timing system, the biology underlying fatigue, interactions of the circadian and homeostatic regulatory systems, and resulting individual differences and performance prediction. The Air Force supports such a program of research with hopes of coming up with information that can be used to develop new strategies to improve performance impaired by jet lag and shift work, night operations, and the loss of life and aircraft because of stress, inattention, or lack of vigilance.

My program includes about 30 grants and contracts to top scientists at well-known universities, in industry, and in Air Force laboratories. The scientific disciplines employed in this research include biochemistry, molecular biology, physiology, and neuroscience, as well as animal and human behavioral studies.

My primary duties and responsibilities include: (1) formulation of program needs and requirements; (2) informing the relevant scientific community about the opportunities available in the program; (3) encouraging submission of grant proposals; (4) evaluation and prioritization of new research investment options; and (5) selection of projects for funding.

Part of this process involves peer review. This could include setting up panels or sending the proposals out to appropriate scientists for individual reviews. I determine which method of peer review and advice is most useful to me and then I implement it. I oversee all technical, fiscal, and administrative aspects of my program. I also am responsible for defending and explaining the program to management and to scientific review boards.

I am also expected to promote and coordinate optimal information exchange between the relevant scientific community and the operational Air Force, including training and education programs, technological development, and application of research findings. Not only is it important that I encourage that these transitions take place, I must clearly document them as evidence of the usefulness of my program to the Air Force.

I regularly coordinate with my counterparts in other funding agencies to assure a research program that meets national policy objectives and avoids duplication. It is also important that I liaison with relevant scientific and technical activities both nationally and internationally, including scientific progress directly within the areas we are funding at this time, and also within related areas, and keeping abreast of progress within my general area of expertise.

How I Got Here

I finished graduate school, taught for a couple of years, completed a postdoc, and then decided that I did not want to spend the next few years getting to know more and more about less and less. I liked the intellectual searching part of research, such as coming up with the research questions to ask, and I liked writing about results and interpreting them for publication. But I hated the actual hands-on experimentation, which I found boring and frustrating.

So what should I do? This was back in the days when every scientist around me, my thesis advisor, my postdoc mentor, and my friends all thought I was crazy at best, a traitor at worst, for wanting something different. After all, I was offered a couple of very respectable academic jobs, why would I want to do anything else? It must be because I was a woman and not really serious about science.

It was a pretty unpleasant experience, coming face-to-face with such strong prejudices. I was not able to find anyone within my academic world who was willing or able to help me explore other options. Somehow, through a long chain of personal inquiries, I was contacted by an Air Force officer at the Air Force Office of Scientific Research (AFOSR), who was looking for a temporary replacement for himself while he went off for a year at the Navy War College.

I didn't even know the Air Force funded basic research at the time. This was the early 1970s, and academe and the military were not the best of buddies. This officer managed a program of research in education and training, and he found out from the folks at the National Academy of Science that I might be interested in trying out such a position, that I was familiar with the scientific discipline critical to his program, and in fact knew several of the scientists funded through the program. I decided to try it for a year. Why not? It would at least give me a year of experience in grant administration.

So in 1979, I started an IPA (an agreement between a university and the government where the government pays your salary and the university agrees to hire you back after your stint with the government). The first day of work they put me on a military airplane and took me to Texas to listen to the three services defending their programs to AFOSR. I'll never forget my reaction when I looked at all of those number charts and realized that the numbers were in millions of dollars. That first year, Jack had already made many of the decisions about who to fund. But our boss was very curious about my opinion of these scientists, and boy, did I have opinions. It was very different talking to brilliant senior scientists as their program manager than it had been from the position of a young, adoring new Ph.D. But we were on the same side, after all, and I enjoyed every new thing I learned.

I put together my first program review of all these scientists, and I went to the annual meeting in my specialty as a very special person. AFOSR decided to offer me a permanent position. I had convinced my boss that the Air Force really needed a distinct program in visual psychophysics, which I was very capable of putting together and managing. And then I gave one of the best briefings of my life to the then-director of the organization, who said immediately after it that they should hire me fast.

And so began my career in the military. I enjoyed learning as much as I could about the Air Force and, for that matter, about the military as a whole. I started with no knowledge at all, but had freedom and support to find out all I could absorb. When I decided that there was nothing I'd like better than to go to War College, I received the support of the Command and was sent to the Industrial College of the Armed Forces (ICAF) for a year.

And what a year that was. The other students were Air Force, Navy, Army, and Marine officers, with a few civilians thrown in for variety. Never have I learned so much in such pleasant and stimulating surroundings. When we studied the Supreme Court, a few of the justices came in to tell us about it. Our studies of banking included visits with senior officers from the top banks in the United States. The general theme of the year was industrial mobilization for war and acquisition within the Department of Defense. These were the Reagan administration years, and the military was being built up. ICAF was a rare opportunity, one not to be missed if you ever get the chance.

After ICAF, I wanted a chance to use my new knowledge in something oriented more toward Air Force than toward a basic research organization, so I took a staff management position at Headquarters, Air Force System Command, as Program Manager in the Directorate of Life Sciences of the Air Force Office of Scientific Research. I became the headquarters representative for all applied research in the area of human systems. In this post, I planned, initiated, developed, and managed programs of basic research in education and training, human visual information processing, artificial intelligence/image understanding, and neuroscience. Each program included 20 to 30 contracts and grants to universities and industries. I formulated program needs, evaluated research proposals, and selected projects for funding, then I oversaw all technical, fiscal, and administrative aspects of the research programs. I defended the programs though the 5-year budget cycle.

Gradually over the next 6 years I was offered and accepted headquarters positions of increasing responsibility, always in some way related to research and development programs. I even spent a year on developmental assignment in the office of the Undersecretary of Defense for Acquisition. But staff work is not really my cup of tea, so when the chance came to return to AFOSR to manage a basic research program, I jumped, even though it was in an entirely different scientific area.

A Typical Day

There is no typical day—there are always new challenges. I have the freedom to structure my time pretty much as I please. Responsibilities shift, depending upon the time of the year (some duties occur annually) and the needs of the moment. Also, my responsibilities depend on my current interests. I interact with a broad variety of people, from the most scientifically sophisticated Nobel Prize laureate to the youngest and most enthusiastic airplane mechanic, the occasional politician, all levels of military war fighters, folks from the financial, contracting, and legal worlds, people from industry, movie directors, and toy makers. Some are very bright and some are not. Some are flexible, and some are rigid.

All in all, I interact with a much greater variety of people than I ever would have if I had stayed in a more traditional scientist's job. This is both exhilarating and a challenge. I cannot expect most of the folks with whom I interact to understand the experiences that shape my thinking and decisions; their backgrounds and interests are very different from mine. I enjoy this, but establishing a common ground for communication is not always easy.

I spend some days at the computer, sending e-mail to my principal investigators (PIs), answering e-mail requests about my program from prospective PIs, designing briefings to explain and defend my program to those above or to try to sell a new program, answering numerous requests for information from a variety of Air Force management layers and oversight groups at the Office of the Secretary of Defense, and sometimes handling requests from Congressional staffers.

Other days I go to meetings. I attend from 2 to 5 scientific conferences a year, some very specific to the subject matter of my current program, and some tangential and/or more general, or even related to new areas about which I'm curious and that I'm considering adding to the focus of our program. Every year I attend a variety of coordination meetings with my counterparts at other funding agencies to find out about their programs and to exchange ideas. Sometimes these meetings are actually turf battles. Of course, I also attend a number of nonscientific meetings. For example, this year, I am the organization's representative to a review board related to the reorganization of the Air Force Laboratories.

Yes, managing a program of research for any of the military branches requires a large amount of travel—to scientific meetings, to university laboratories, to Air Force operational bases, and to Air Force research and development facilities. Many of the program managers here at AFOSR travel 2 or 3 times per month; some of this is discretionary and some is required travel.

And most important, it is a requirement that I present my program at the meetings of a number of review boards—sometimes scientific, sometimes operational, sometimes both. The purpose of these meetings may be to determine whether the Air Force will continue to fund research in this area, or it may be to convince the Department of Defense (DOD) that the three military services have complementary rather than duplicative programs in an area.

I also organize my own meetings. Once a year or so, I have my grantees present their research to a review board I've set up to evaluate the program on the basis of the science, as well as for possible long-term Air Force application. Twice a year, I hold a working group composed of scientists and Air Force operational types to exchange information. I want the scientists to develop an in-depth understanding of relevant Air Force issues, and I encourage the operational folk to take advantage of advice to be gained from the brilliant minds of my researchers. It is also possible that I will put together a conference about a new area we are thinking of developing.

Some days I read proposals and proposal reviews, make decisions about who to fund, and then write the documentation necessary for our contracting people to write a grant. Some days I read scientific journals and textbooks. Other days I spend the whole day on the telephone, putting out fires, keeping in touch, making arrangements.

There are also other projects that arise because of current management interest in the topic. For example, this year our organization is very concerned with expenditure rates on grants. Expenditure is defined as the point at which the Defense Accounting and Finance System actually cuts the check in response to an invoice sent in by the grantee's institution. An understanding of this process was not part of my job before, but now it is. So I investigate. I track down how the process works, where the kinks are, who has the relevant information, who can fix the problems, and how I get to them. I'd hate to have to spend a whole lot of time on these financial details, but it is interesting to come to understand it and to solve the problems of concern to my management.

A couple of years ago, I was selected by the Director of AFOSR to investigate and improve our agency's relationships, interactions, and image in all Air Force Laboratories. The first year, I visited all the Air Force Labs and made myself available for personal and/or small group discussions with as many of the scientists and managers as possible. I met a lot of people from a number of disciplines and I listened and took notes. I learned a tremendous amount of science and application, and management and interpersonal issues. I wrote up and presented my findings and recommendations to senior management at all the labs and to management and all the PMs at AFOSR.

AFOSR implemented my recommendations and our relationships with the labs have never been better. And I found myself with the additional duty of visiting these labs every year to make myself available to listen to problems.

REQUIRED SKILLS

You need to be analytical, creative, able to handle interpersonal interactions, able to deal with computers and applications, and more. You bring what you are as well as who you are to this, as to any job, and you use all of your talents and skills. What am I good at? Well, I think I am a very good listener. And I'm pretty good at seeing the world from another person's point of view. This helps me figure out the best examples and the most appropriate kinds of arguments to convey my point. I am extremely curious about new things and I like change and challenges. Repetition bores me and when I am bored, I get mean. I am a pretty good briefer, especially if I have time to really prepare. I am able to take risks, make decisions, and live with the consequences.

TRAINING

This is on-the-job training. I am always training and learning new things, and I started more-or-less productive work my first day on the job. The Air Force is very generous about providing both part-time and full-time continuing education. I've been to War College, I have almost enough credits for an MBA, and I am able to attend scientific and management workshops and short courses pretty much as often as I like. Other program managers have taken sabbaticals to spend a year doing full-time research, to write books, and to become involved in diverse special projects.

JOB STRUCTURE AND SALARY

Government salary structure is based primarily on the GS (Government Service) rating for a position. I make just under $100,000 as a top-level GS-15, including bonuses. Starting salary in government for a new program manager will depend somewhat on experience and academic discipline, but I think we occasionally hire at the GS-12 level, which makes $40,000 to $50,000 annually. Top salary for higher level managers, Senior Executive Service, in my chain is from $120,000 to $130,000, but the nature of their job is completely different; they manage the managers of science.

There are many opportunities for advancement and for lateral job moves, but none are in program management. For example, my boss moved from program manager (PM) to director of the Directorate of Chemistry and Life Sciences. He has 7 PMs working for him, and we manage the programs while he manages us. His job description reads as follows: Plan, direct, evaluate and coordinate the $50 million Air Force basic research program in the fields of chemistry and life sciences, encompassing such highly specialized areas as organic, inorganic, and theoretical chemistry and physiology, toxicology, biochemistry, molecular biology, neuroscience, biophysics, and human engineering. Responsible for all operational, policy, and programmatic activities of the directorate, including all United States Air Force basic research in these areas, about 250 grants to AF laboratories, industry, and academia. Ensure that scientific merit is high and that goals are relevant to long-term Air Force needs. Encourage familiarity and promote appropriate integration of basic research into Air Force and other DOD applied research and development programs. First level supervisor/leader to Directorate Program Managers, all highly respected senior level scientists.

I have also held a number of other "lateral" government positions, some as training assignments and some because I actually changed jobs. These positions were primarily headquarters-type staff management positions that included developing policy and defending budgets, and staff management of large programs of research. For example, in 1984 I transferred to Headquarters, Air Force Systems Command, at Andrews Air Force Base in Maryland, where I spent 6 years in a variety of staff positions with increasing levels of responsibility and authority.

Headquarters staffers spend most of their time explaining and defending programs throughout the DOD and Congressional budget cycles. I was responsible for a variety of basic and applied research and development programs, all vaguely related to the scientific disciplines I studied as a graduate student and a postdoc.

I also spent about a year in the Pentagon working in Program Integration and Strategic Planning for the Undersecretary of Defense for Acquisition. This was a fun job. The office was a small, very tightly knit group of people, including one Air Force officer, one Navy officer, one Army officer, one Defense Intelligence guy, one Central Intelligence guy, one civilian who came from the National Security Council, and me.

Half of us had scientific and technical backgrounds, half of us had strong history and political science backgrounds. Our role was to integrate information about world events and develop both short- and long-range Department of Defense acquisition policy. We represented the Undersecretary in situations where national policy questions impact military acquisition programs in more than one specific technical area. For example, we integrated classified and open source inputs and we wrote the technology sections of President Bush's first major National Security Review. I also developed the congressionally mandated plan to use Department of Defense technological resources to help implement the President's national strategy to control the use of illegal drugs. It was great fun, and I learned so much.

WHERE CAN YOU GO FROM HERE?

There are a number of possibilities in industry, in academia, and in other government or quasi-government agencies. Large industries that deal with the government are often eager to hire an individual whose knowledge and contacts might help them. I am not referring to anything dishonest here. Companies are eager to have on board folks who know their way around the military acquisition of science and technology, and of systems.

There are also opportunities in the so-called "beltway bandits," or think tanks around Washington. Many of the military folk who retired from AFOSR have taken university positions such as dean of research or provost.

A few have actually gone back to university teaching and research positions. There are always opportunities within other government agencies; I've seen advertisements from the General Accounting Office, the Office Of Management and Budgeting, and from the Library of Congress, and even the Smithsonian Institute, for which I have exactly the experience requested. It would also be possible, and very interesting, to become a congressional committee staffer. There is a government program for young and eager employees of the executive branch to work "on the Hill" for a year or two.

What Opportunities Exist in the Military?

The United States Air Force is a large organization, and only a very small proportion of it is dedicated to science. The Air Force Office of Scientific Research is responsible for all of the basic research supported by the Air Force, whether at a university, within industry, or in a government laboratory. At the moment we have about 40 program managers. We seldom employ more than one scientist from a particular specialty area. Areas in which we currently have research programs include: structural materials, mechanics of materials, particulate mechanics, external aerodynamics and hypersonics, turbulence and internal flows, air breathing combustion, space power and propulsion, metallic structural materials, ceramics and nonmetallic structural materials, organic matrix composites, electromagnetic devices, novel electronic components, optoelectronic information processing, quantum electronic solids, semiconductor metals, electromagnetic materials, photonic physics, plasma physics, imaging physics, chemical reactivity and synthesis, polymer chemistry, surface science, molecular dynamics, chronology and neural adaptation, perception and cognition, sensory systems, bioenvironmental science, dynamics and control, physical mathematics and applied analysis, computational mathematics, optimization and discrete mathematics, signal processing, probability and statistics, software and systems, artificial intelligence, electromagnetics, meteorology, and space sciences. These areas do change from year to year and can be found listed on the Web at http://www.afosr.sr.mil.

Skills Needed to Succeed

Communication skills are absolutely essential for the effective program manager. I regularly explain and defend my program to higher level scientific management and scientific review boards, and to a number of different nonscientific groups, in particular and most importantly to Air Force operational types, who have no time and just want to know the bottom line—what the research is good for and how it will help them. I also frequently have to explain program needs to the scientific communities that are involved, both in writing and verbally to individuals and to groups.

A wide range of personalities can do well in this setting. I'm an introvert and I do this job well and enjoy it very much. But I hardly think being an introvert is required. I suspect qualities such as independence, integrity, humor, patience, consistency, clarity of thought and communication, curiosity, and decisiveness are important. After thinking about this question for a while, I asked other program managers in my organization for their thoughts. The following is a sample of their responses:

1. A strong interest and training in a scientific discipline.

2. A willingness to sacrifice the satisfaction of pursuing your own special research interests and ideas in that discipline.

3. The ability to tolerate and adapt to inconsistencies and frequent changes in administrative policies.

4. The analytical ability to understand the bottom line of complex scientific issues.

5. The ability to express the bottom line clearly in oral or written format to others who may not have your technical training.

6. The ability to make decisions and execute actions based on those decisions.

7. The curiosity to explore and understand scientific issues outside of one's specific discipline.

8. The ability and inclination to interact with a broad range of scientists and administrators on issues of importance to them.

You must believe in your goals and take a position of strong advocacy to defend them.

THE PROS AND CONS

I most like being able to significantly influence the course of science—not simply through one discovery or one research project, but rather by directing a whole program of research in the direction I believe is most profitable.

I least like the many and continuous bureaucratic requirements that are repetitive and boring, and that distract me from my scientific program.

The Department of Defense is a huge bureaucracy, and all of the niggling red tape common to any organization is certainly present here.

In coming to the military in the 1970s as I did, I never expected to find so many of the military officers to be highly intelligent, creative, well-read, cultured, sophisticated, and fun. Academics, at least when I was still in school, were convinced that the best and the brightest stayed in academia. I was pleasantly surprised and delighted to find that this is not the case.

But the two cultures are very different. At first I did not enjoy having to dress for work—I wore jeans at the university, even when I taught. More seriously, academics are rewarded for questioning authority. In fact, a student has not really made it until he or she disproves the major finding of his mentor. Well, the military is not like that. Loyalty is probably the most valued quality in this environment. I thought I was showing that I cared by attacking all the weaknesses in the arguments of my government lab scientists. They were absolutely crushed and they thought I was a terrible person who was about to destroy their projects. This is a real cultural adjustment, which I have only partly managed to achieve. I still question all authority.

I love "hair on fire" days and I love stress; there are not enough of those days for me. I do not like frustration, and any government bureaucracy has plenty of frustration. There are too many of those days.

There is a world of difference between academia and this position. In graduate school and as a postdoc, I did hands-on research—I made electrodes, stayed up all night with sick cats and sick computers, wrote proposals, journal articles, and scientific presentations on a very esoteric topic for a small, informed audience of scientists in the same or in a very similar esoteric field. My time was my own, dependent only on the requirements of the on-going study.

Now, I stay up all night preparing briefings or participating in budget exercises. I also need to respond to and be responsible for other people. The types of people with whom I must interact are more diverse and the communications process is a much more creative and exciting challenge. I have to translate my world into language that they understand, and I have to get them to care about my issues.

How to Get This Job

The best job listings are found in *Science,* and usually in the major journal of the scientific discipline being requested. *Commerce Business Daily* is also a good source. The key is to begin networking right now, if you want to really learn what is out there.

The government is a maze, and it can be very difficult to navigate without insider help. I believe that most government jobs are advertised on the Web (www.fedworld.gov/pub/jobs and www.USA jobs.opm.gov/a3.htm). There also are a number of fellowships available to get you in the door for a year or two so that you can learn about the structure and the culture, and decide if this is really for you.

The AAAS Defense Fellowship is a new option. This year, we awarded two of these fellowships, hiring a person for the Acquisition Office of the Secretary of Defense, and hiring another person to assist the Deputy Assistant Secretary of the Air Force for Science, Technology, and Engineering.

The needed qualifications are a moving target. We seem to hire many more senior scientists now than we did when I was hired. I suspect that doing good science is the most important criterion. I would not be able to do my job (nor would I have been hired) without having had excellent scientific training. Most importantly, my training and experience doing science taught me how to recognize good science. It also taught me how to talk to scientists, what to look for, how to start researching a new area, and how to understand whole science culture.

Because communication with a diverse population is crucial to performing the job well, it is critical to be articulate and to have the ability to explain science to the layman during the interview process. We do still hire IPAs (academic contracts) for 1-year assignments. And it is possible that it would help to become known as an Air Force laboratory bench scientist. There are 1-year government fellowships designed to give people a familiarity with government. One could also be trained in program management by taking a rotator position at the National Science Foundation.

23

BUSINESS INFORMATION SERVICES:

Providing the Data for Industry

Mark D. Dibner, Ph.D., MBA
President, Institute for Biotechnology Information

I could not have planned my career in advance, as the biotechnology industry did not exist when I was getting my undergraduate degree, and it was
just barely getting off the ground when I completed my Ph.D. I thought
I was going to wind up as an academic scientist for life. Instead, I wound
up in a large corporation, and the evening MBA program I thought I was
going to hate actually turned out to be enjoyable. An opportunity I never
expected materialized, causing me to leave the research lab for good to
enter a career in business information a dozen years ago. Today, I have two
"professions" that I work at each day—I am in the information business and
I am an entrepreneur.

HOW MY PATH CHANGED DIRECTION

I was always sure that I would end up as a physician—it was my goal. In
high school, I took all the science classes I could cram into my schedule.

Even my extracurricular activities were geared toward medicine—I was in an Explorer Scout Medical Post that met with different types of doctors each month and manned the first aid tent at all the scout jamborees.

Following high school, I majored in psychology and premed at the University of Pennsylvania, where I matriculated in 1969. As a psychology major, I shunned the "touchy-feely" psych courses such as social psychology, and felt more at home in a very strong physiological psychology department. The best professor I had there, Dr. Philip Teitelbaum, had done some of the original work demonstrating that specific parts of the brain controlled specific bodily functions. He showed that stimulating a particular area of a rat's brain caused that rat to get extremely fat and insatiably hungry. Or, he could cut out that same part of the brain and the rat would get very thin, seemingly lacking any appetite. Dr. Teitelbaum had lots of theories that he was pursuing.

I had a theory of my own, which led to my first scientific research project having something to do with plying college freshmen with alcohol mixed with Hawaiian Punch and checking their handwriting. I have no idea how this research got past the Human Subjects Committee (the freshmen were under the legal drinking age) and a member of the committee even gave me a gallon of the ethanol from his lab to get me started. That was the birth of my scientific career. This research in physiological psychology solidified my major in this subject and my conviction that I would pursue a research career.

Upon graduation, I enrolled in Cornell University's Graduate School of Medical Sciences, entering a Ph.D. program in neurobiology and behavior. This school was part of Cornell University Medical College in New York City. Early on, I sought out a Ph.D. advisor within my interest in developmental neuroscience. Dr. Ira B. Black had never had a graduate student before and was just finishing up his residency in neurology. But he came with a good pedigree. Following the receipt of his M.D. at Harvard Medical School and an internship, he took time to pursue research at the NIH with Julius Axelrod, a Nobel Laureate for his work in catecholamine research. Ira had also traveled to Cambridge, England, to spend a year in the lab of Leslie Iversen, another Axelrod disciple.

I spent nearly 4 years in Ira's lab, studying the effect of target organs on sympathetic neuron development, that is, how the things that nerves grow to and attach to affect the development of those nerves reaching them. Although I was Ira's first Ph.D. student, he had clear thoughts on what he wanted to teach me and how I could progress through a strong research regimen. In succession, he taught me to think like a scientist (e.g., never make statements that were not properly supported by fact or empirical knowledge), how to design experiments properly, how to write well (he was a merciless editor), how to write grants (not only did I help with preparation of lab grant proposals, I won an individual NIH predoctoral fellowship), and how to make presentations.

There is some question about the value of the Ph.D., especially for people who choose not to use it. To me, the value I feel for myself, and see in other Ph.D.s who work with me, is the ability to have taken a large body of work to completion. This includes becoming the world's expert in one specific field and creating new understanding or knowledge in that field. People who have gone through that exercise have the ability to focus on projects, to design and implement the solution, and they are able to write and speak about it.

Thanks to Ira's focus, I was the first Cornell Graduate School of Medical Sciences Ph.D. student to make it through in only 4 years. A postdoctoral position was lined up in the laboratory of Dr. Perry Molinoff at the Department of Pharmacology at the University of Colorado Medical Center in Denver. Perry was another Axelrod disciple. My grant writing skills allowed me to line up both NSF and NIH postdoctoral fellowships. Although the mentoring was minimal, Molinoff did pull together some other excellent postdocs, and we worked on adrenergic receptor pharmacology and the regulation of these receptors.

After two-and-one-half years in the Molinoff Lab, I left for another postdoc/junior faculty position at the Department of Pharmacology at the University of California at San Diego, working with Dr. Paul Insel. At this lab, I picked up skills in cell biology and again used my grant writing skills to work on lab grant proposals and get fellowships, including a California Heart Association fellowship. I stayed in San Diego for 14 months.

There are two things I want to point out about the postdoctoral years. First of all, I firmly believe that most scientists experience tremendous personal growth and do their best and most creative work during their postdoctoral years. Their future scientific careers are based around what they did during their postdoc, much more than what they did for their Ph.D. research. They are in much more of a doing rather than a learning mode and they can be much more productive. They work between 60 and 80 hours a week, churning out research and papers at a strong pace.

These are tough, but productive and important years. They are important for the development of a career, either in science or in other career alternatives that one may choose. Critical thinking, writing, and presentation skills are honed during these years. We also learn that we can work hard, over long hours, and survive. The postdoc, like a medical residency, is an important learning experience.

CAPTAIN OF INDUSTRY

I did not want to go into industry. I was supposed to be an academic scientist following in the footsteps of my mentors. In the mid-to-late 1970s, nobody I knew went into industry and there were just a handful of biotech firms. But fate stepped in. E.I. Du Pont de Nemours and Co. had decided, in 1980, to go into the life sciences to diversify the company beyond petrochemicals for its bright future. Du Pont hired the best think tanks to figure out what it should do for its future, and they generated stacks of reports pointing to the life sciences and specifying about 10 areas in which Du Pont should focus. One was neurobiology. The company proposed to spend many millions of dollars to build its life science capabilities.

I had put a brief résumé together and sent it to FASEB for their job search service. My background, field of interest, and strong publications list fit the profile of what Du Pont was looking for in building their neuroscience program. They called me and asked to interview me at the upcoming FASEB meeting. I told them I had no interest in industry. Indeed, I had checked the box on the FASEB form that said I was not interested in industry. But, they asked to be heard. To be honest, I knew little about industry or industrial scientists—I just knew that I was destined to be an academic researcher. But I agreed to meet the Du Pont representatives anyway.

What they had to offer was too good to be true. They were inviting me to join their Central Research and Development Department (CR&D), the basic research arm of Du Pont, set up originally in the 1920s with the idea that if you put enough bright scientists together you would generate good ideas that would benefit the company in the long run. CR&D generated nylon and kevlar, among other things. Now it was going to generate good life science ideas. Not only would I be doing basic research, but they had earmarked an average of $275,000 per year per scientist and I would never have to write a grant to get it. The salary, while not as high as I had imagined, was about 20% higher than I was seeing in academic departments for an assistant professorship. I signed on Du Pont's dotted line.

In 6 years at Du Pont, I was never told what to do in my lab. I had academic freedom, I could publish papers on my work (with a few restrictions), I had the opportunity to travel, and I enjoyed the company of a wonderful crew of bright young scientists they/we were pulling together.

Many of my ex-colleagues, still in academia, accused me of "selling out to industry." They actually used that phrase, and often with contempt. You see, they knew as little about industry as I did, but they, too, felt that to be pure, one must stay within the ivory tower. However, after about a 5-minute explanation of the complete freedom, the huge financial support, and the lack of grant writing, many of them asked me if we still had openings. It wasn't until the late 1980s, when the number of biotech firms became significant, that the biological science community begin to understand that there were opportunities in industry.

In my 6 years at Du Pont (1980-1986), I was able to do good research. One of the highlights was that, after only one year there, they allowed me to do a 4-month sabbatical in Cambridge, England, at Leslie Iversen's laboratory, following in my Ph.D. advisor's footsteps and working with some bright young scientists. Other skills were picked up in industry, especially an appreciation and understanding of intellectual property, some personnel management related to the 2 or 3 employees who worked in my lab, some human resource skills such as the interview and hiring process, and an understanding of how to work within—and sometimes around—a large bureaucracy.

After 2 years at Du Pont, I found out that the company would pay for advanced coursework if it related to my job. Du Pont took a broad view of related coursework and preferred it to be related to a degree program, not just isolated courses. This generally meant a law degree or an MBA. I was single at the time and had liked school in the past, so I wanted to fill in some of my evenings doing something. I knew I would not like law school and I had no interest in business. But the MBA program was the lesser of two evils. I enrolled in Widener in 1983 for their evening MBA program.

Since I did not think I would like business school, I decided my first course should be the one I was convinced I would hate the most. The plan was to hate the course so much I would get this coursework idea out of my system and drop out. Good plan. So, I took Managerial Accounting. I did not know a debit from a credit (still don't) and felt it was time to learn this. It turns out that this course was in fact about strategic decision making. It included two of the most useful concepts I learned during the MBA program –sunk costs and opportunity costs. I actually liked business school!

I stayed with the MBA program for the ensuing two-and-one-half years. I was doing good, but not earth-shattering, research at Du Pont while finding I enjoyed learning about strategic planning. Also, Widener had a requirement that would drastically alter my career—they required a Master's thesis of all MBA students. I chose to write about the impact of biotechnology on the pharmaceutical industry.

Being a Ph.D., I wasn't about to write a thesis-length paper without publishing it. I had created two databases, one on U.S. biotechnology companies and one on the actions and alliances that biotechnology companies were forming worldwide, as part of my background work for the thesis. I published two papers on the advent of the U.S. biotechnology industry.

Stuff happened. Good stuff. The editor of *Science* phoned and asked me to write a lead article on U.S. versus Japanese commercial biotechnology. I was called by some international society (I forget which one) and asked to spend a week in the Banff Springs Hotel and give a keynote lecture at their annual meeting. I received a phone call from the Science Indicators Unit at the National Science Foundation and was asked if they could give me money to do additional studies of the U.S. biotechnology industry for their *Science Indicators Report*. The head of the Wharton School's Management and Technology program called and asked if I could work with him part-time as a Fellow to continue to build the databases.

As a research scientist I had never been asked to write a lead article for *Science*, give a keynote speech, get unsolicited NSF funds OR become a Fellow of a top school. I had stumbled onto something. I had found a niche. And it was something I liked. A career was born.

At first, I wanted to use my brand-new MBA in strategic planning to help the Du Pont company reach new heights—after all, they paid for it. But as it turned out, Du Pont did not have any mechanism for taking people who have gained new skills, even skills the company had paid for, and using them for the company's benefit. I was able to land some internal consulting positions, but was told that they had hired me to be a scientist so it was not going to be easy to switch me to a different career path.

Then the unimaginable happened, which made me realize I had to leave Du Pont. They were forming a new group, a strategic planning group for biotechnology. Not only did I have my brand-new strategic planning MBA, but I was becoming known as an expert in biotechnology (not only from my industry studies, but I was also finishing, with two coauthors, the writing of a textbook on molecular biology techniques). Maybe, I thought, I could even head this new group. Du Pont had other ideas. They chose five people, chemists and engineers from within the company, none of whom had background in either biotechnology or strategic planning. I had to get out.

Adding icing to this cake were two pieces of advice given to me by a Du Pont lifer with whom I was asked to complete a special information project. When I wanted to make the database we were creating highly sophisticated and useful, he said "Never give them more than what they ask for." When I wanted to let other groups at Du Pont know about the availability of this database, he said "Never stick your head above water and you wont get shot." I knew he was right—for this corporate culture—and it meant that I had to leave the company. Also, I was ready to give up my career as a research scientist for something new and exciting.

Creating a New Career: The Proactive Approach

During the Christmas holiday of 1985, while still a Du Pont researcher with a 5-month-old MBA, I sat down and did something hokey. I wrote down on paper the answer to this question—If I could create a job that would be exactly what I wanted to do, what would it be? In my answer to myself, I described creating a group that would study the biotechnology industry and provide strategic business information. This group would have three main resources—the databases I had begun for my MBA thesis, a good library of resources specific to commercial biotechnology, and staff who focused only on commercial biotechnology. Given enough time and financial support to build these resources, I suggested that the group could provide strategic information services for a fee.

The next step was to use the miracle of word processing to take this virtual job description and turn it into a letter. I had interviewed a new type of organization—biotechnology centers—for my *Science* article. In 1985, there were only two centers of any repute, the Maryland Biotechnology Institute and the North Carolina Biotechnology Center (NCBC).

I wrote a letter to each of them telling them that if they would invest in me and my Biotechnology Information Program as a new division of their centers, then the Program would bring them international attention, provide a great resource to the community they serve, and ultimately bring in some funding for the research we could do with the possibility of someday becoming self-sufficient. As it turns out, the Maryland Biotechnology Institute was mostly creative publicity at the time, but NCBC was real and very eager to talk. They had only eight employees at the time and more money than they could spend. Timing is everything.

We talked, my wife and I moved to Durham, and the Biotechnology Information Program was born at NCBC. I left for-profit, large industry for a small, state-funded, private, not-for-profit organization. In retrospect, all of the things I put in my letter selling the program to them have come to pass—and more. The Program evolved into the Institute for Biotechnology Information in 1992 to give it an aura of a separate entity, even though it was still a division of the Center. I was director of IBI and a vice president of the Center.

We had 9 employees in IBI and an international reputation. The databases grew tremendously and we sold them. We collected data and published them as directories. We were asked by companies to do special studies for them. And we were asked to serve as government contractors and subcontractors to complete studies. The library had grown to be what was arguably the best public library in the world for commercial biotechnology. The biotechnology center got the attention it sought, along with some income, and was able to provide a tremendous resource to the community it served.

While at NCBC, I was able to pursue four other things I enjoy very much. First, I was able to teach. As another example of timing, my moving to North Carolina coincided with the firing of a nonproductive faculty member of

Duke University's Fuqua School of Business. I met with the Dean to see if I could teach perhaps an occasional lecture and walked out an Adjunct Associate Professor teaching one or two courses each year in management of technology and, more recently, entrepreneurship. I have been with Duke for 12 years now. I had previously taught courses in Denver and San Diego during my postdocs, but had never taught graduate students. My fellowship at Wharton and my lead articles in *Science* led them to believe I knew what I was doing.

The second thing I was able to do at NCBC was to continue to write and publish. While there, I authored 4 books, 2 on U.S. biotechnology and 2 on Japanese biotechnology. We were doing numerous studies and I was able to publish about 50 papers during those years, mostly on commercial biotechnology. Working for NCBC, I did not have to account for my hours to be financially productive, so I could take the studies we were doing and follow through with published articles.

The third thing was to continue to travel and make presentations about my work. I lecture about 15 times each year at a national or international venue. I enjoy lecturing and it helped me spread the word about our work to thousands of people each year. Like the publishing, it was a form of free marketing, and getting out and about was great for networking.

Networking is the fourth important skill I was able to build during my years at NCBC. My boss, Dr. Charles Hamner, President of NCBC, was an excellent mentor. He was happy to have us spread the word about NCBC at meetings, with societies, wherever. I joined the Board of Directors of the Association of Biotechnology Companies. I founded the national Council of Biotechnology Centers and served as its chairman, I got involved with the Drug Information Association, running the Biotechnology Track at their annual meeting, I gave keynote lectures to countless societies and groups, I served on National Research Council Working Groups, I was on three editorial boards of industry journals, I was on the boards of directors of two companies. I networked.

During my eight years at NCBC, our IBI group also grew and established many ties. We worked for an increasing number of clients. We created more than 20 published books and directories. We were quoted frequently. Our abilities to do market research, competitor analysis, technology assessments, international studies and a panoply of special research projects grew. This was a team effort of people with great abilities.

It was nice working for someone else who was supportive and nurturing, as both the Center and Dr. Hamner were. But IBI was growing into its own entity and approaching self-sufficiency. The state legislature also realized this. They saw that NCBC had a division that was actually bringing in money.

At the beginning of 1993, they gave Dr. Hamner a mandate to make IBI self-sufficient within 5 years. He, in turn, gave me 3 years to accomplish this goal. We talked and agreed that if he gave me 1 year to plan and prepare, I would take IBI as a private, for-profit company. This had benefits all around—the Center got to beat the mandate from the legislature by 4 years, it got to spin off a company for the first time, which would help satisfy its mandate of creating jobs and revenues related to biotechnology. I had confidence that IBI could succeed as a company. I was ready to become an entrepreneur.

ENTREPRENEURSHIP

The Institute for Biotechnology Information, LLC was launched in July 1994 as a new company owned by my wife and myself (and financed through remortgaging our house). We took three people from our group at NCBC and hired two others to begin the new company. A contract was drafted giving IBI 3 years' of work from NCBC to continue to provide some of the information services we had been supplying internally. We had to leave the library behind, as it was a showplace and attracted considerable attention. But the Center was bound to maintain the library for 5 years and could not compete in providing information services for profit for that time. IBI moved to new, modest digs nearby in Research Triangle Park.

IBI provides strategic business information in biotechnology and pharmaceuticals to a wide variety of clients, from government agencies to biotechnology firms to large corporations. About one-fifth of our income comes from sales of products—four books, a monthly journal, and our databases (if the reader is interested, details can be found at http://www.biotechinfo.com).

Four-fifths of our income comes from special studies, which vary widely in nature and scope. For example, at the time of this writing, active projects include a market research study in Egypt, creating a database for a client, writing a complete business plan for a new company, a market research study, an international data collection project, collecting financial information on selected companies, and a few others.

We have also begun to focus in areas of pharmacoeconomics (as a joint project with the world leader in this field), regional development (we work for states and regions to help plan their development of commercial biotechnology) and business plan writing.

IBI has grown in its first 3 years from 6 to 14 employees. We are growing slowly but steadily. The three strengths we have are our resources, such as internal and external books and databases, our reputation and networking, and our people. IBI was created to be a team of people with complementary skills who could work well as a team. The majority of our staff has advanced graduate degrees. Three of us have Ph.D.'s, two have MBAs, and others have an assortment of Masters degrees.

What we provide, in a nutshell, is strategic business information. The main strengths we bring are the strengths I described in my lengthy sojourn through my career. They are the ability to design scientific research, the ability to write and edit, the ability to teach (clients), and the ability to network for both new business development and information gathering.

With the excellent level of personnel availability in the Research Triangle Park area of North Carolina, we could fill most of our positions with Ph.D.'s. In actuality, only the research jobs really benefit from a Ph.D. background, so we make efforts to not hire Ph.D.'s for positions in which they would be overqualified (and never satisfied). Within IBI, the Ph.D. is useful in data gathering, report writing, scientific know-how for a variety of studies we do, and data analysis and assessment.

WHERE THE TIME GOES

I have two jobs—running a small but growing business and working in the information business. My day is almost equally divided between the two. Half of my time is spent nurturing and growing the business. A wide variety of tasks is devoted to maintaining the business. I work on personnel issues, marketing, quality assurance, financial issues, and many other business functions. And managing.

The most important statement on managing is also the simplest. As Harold Geneen, the CEO of ITT put it, "Managers must manage." If you realize that the buck stops here and are willing to make the decisions necessary to keep things running, that is half the battle. But the other half is that management decisions need to be made and you cannot abdicate this responsibility. Of course, early on in a new company much time is spent making some pretty mundane decisions—insurance, telephone systems, copiers, software packages, printers, etc., etc. There is also considerable time spent with a variety of service providers, such as our attorney, accountant, and others.

IBI has been at about break-even level, perhaps a little profitable, during its first 3 years of existence. This is especially good since we have been experiencing continuous growth with the cost of new employees in both time and expense. What I enjoy most is the managing—making the key decisions as they come up. What I enjoy least is personnel issues, especially going through the interview process for new hires.

The task I enjoy second most surprises me. I am involved in all areas of marketing for IBI. This is mostly because potential clients have seen me lecture, or have read one of my books or papers, or have been told to contact me. I am trying to change this, but to date, most new work comes directly through me. I enjoy talking with prospective clients, preparing proposals, and reeling them in.

My other job, as an information provider, is equally challenging. IBI has an excellent staff and we work well as a team, but I have my own role on the team. My areas of expertise, in strategic planning, regional development, and international competitiveness in biotechnology, require that I take the lead in some of IBI's studies. So, for example, when we were hired to do projects for the states of Connecticut and Rhode Island to assist with their local development of biotechnology. In other cases, I am specifically requested to work on a project. Finally, I like to serve as quality control for our studies, doing a final check and edit of all reports that go out.

I typically work long hours, but on my own terms. I have a wonderful and understanding spouse—no entrepreneur should be without one—and a growing son. I work full days, weekdays, but try to get home to my family at a reasonable hour. When the family goes to bed, I stay up and work an additional 2 to 4 hours. I am a night owl and do my best work late at night. This way, I can spend good family time, I can coach soccer or take tae kwon do with my son, and I can still put in the 55 to 65 hours it takes to run a business and get the work done. What does keep me away from the family is the travel. I typically go somewhere one or two times each month. Within the last year, my work has also taken me to Canada, South Korea, Hungary, and Switzerland.

WHERE IBI IS GOING

We continue to experience growth in demand for our services. This is likely due to increased marketing and more familiarity with IBI and its capabilities. IBI will likely grow to between 20 and 25 employees and reach $2 to $3 million in annual sales over the next 5 years. I would like to manage steady growth. This growth will be achieved through gaining additional clients and working on obtaining federal contracts.

Although half of our work to date is as a contractor or subcontractor with federal or state agencies, for the most part we have never initiated contact with these agencies. We have been sought as a sole provider. The next step is to actually try to get new government work *de novo*. For our industry clients, we have a new program called Virtual Information Service, where we can serve as the virtual information arm of small- to mid-size biotechnology firms. This program has proved to be quite popular and is growing daily.

I also have the entrepreneurial bug. I enjoyed starting IBI. I am working on a new joint venture company to provide a needed service to the pharmaceutical industry. This company would dovetail with IBI's efforts and create value in its own right. We'll see.

GETTING INTO THE BUSINESS

We have a variety of jobs at IBI, but the one most relevant to Ph.D.'s in science is on our research team. We seek people with strong research backgrounds who have made the decision to leave the lab. A strong and broad understanding of research and the biological sciences is required. Also required are: an excellent telephone presence for the telephone interviews for primary data collection; excellent writing and editing skills; advanced use of word processing, spreadsheet, and database programs; some statistical and analytical background; and great teamwork. Desired, but not usually available, is a background with business issues and concepts. The business aspects usually are learned on the job.

Two of IBI's researchers hold Ph.D.'s. One has a background in plant agricultural biotechnology, but she has headed dozens of research projects for us in fields from pharmaceuticals to international assessments in her 4 years with IBI. The other has a Ph.D. in molecular biology and she, too, has worked on a wide variety of projects. The other important attribute for a researcher in this area is a high degree of flexibility. We never know who will hire us to research what from one week to the next.

PATHS TO THIS CAREER

Large corporations often have many employees involved in competitive intelligence or in strategic information. Many of these employees come from the research ranks within the company. Other companies, such as venture capital firms or investment houses, need people with scientific training to assess proposals, markets, and competition in order to make sound investment decisions.

Unfortunately, most Ph.D. scientists do not have the business background necessary to work in these areas, nor do they tend to have any sort of formal information training, although with Internet searching becoming more ubiquitous, this training may be obtained on a de facto basis. The business training can be received on the job if the work environment is conducive to this.

There are not many companies like IBI, but there are some large consulting companies, such as KPMG Peat Marwick or Ernst and Young, that would hire Ph.D. researchers to do business studies. An understanding of business would be helpful, and it can come from experience in industry or even an added MBA.

CONCLUDING THOUGHTS

I could not have chosen this career path if I had tried. Paying attention to unexpected opportunites was important. I enjoy all aspects of the current phase of my career, from the entrepreneurial pressures to management requirements. I have a constant sense of being challenged, and a constant sense of having a growing and highly satisfying career. The work always takes exciting twists; we get in new research projects weekly and we have the chance to learn about many fascinating subjects. The studies we do become good fodder for writing papers and lecturing, both of which I enjoy. Most of all, when we see our market research or business plan lead to getting needed funding for an entrepreneur, or when we pull together information that is used to further commercial biotechnology in a state, the personal satisfaction is tremendous.

INDEX